W9-AJL-093

FREE Test Taking Tips DVD Offer

To help us better serve you, we have developed a Test Taking Tips DVD that we would like to give you for FREE. **This DVD covers world-class test taking tips that you can use to be even more successful when you are taking your test.**

All that we ask is that you email us your feedback about your study guide. Please let us know what you thought about it – whether that is good, bad or indifferent.

To get your **FREE Test Taking Tips DVD**, email freedvd@studyguideteam.com with "FREE DVD" in the subject line and the following information in the body of the email:

   a. The title of your study guide.

   b. Your product rating on a scale of 1-5, with 5 being the highest rating.

   c. Your feedback about the study guide. What did you think of it?

   d. Your full name and shipping address to send your free DVD.

If you have any questions or concerns, please don't hesitate to contact us at freedvd@studyguideteam.com.

Thanks again!

Civil Service Exam Study Guide 2019 & 2020

Civil Service Exam Book and Practice Test Questions for the Civil Service Exams (Police Officer, Clerical, Firefighter, etc.)

Test Prep Books

Table of Contents

Quick Overview

As you draw closer to taking your exam, effective preparation becomes more and more important. Thankfully, you have this study guide to help you get ready. Use this guide to help keep your studying on track and refer to it often.

This study guide contains several key sections that will help you be successful on your exam. The guide contains tips for what you should do the night before and the day of the test. Also included are test-taking tips. Knowing the right information is not always enough. Many well-prepared test takers struggle with exams. These tips will help equip you to accurately read, assess, and answer test questions.

A large part of the guide is devoted to showing you what content to expect on the exam and to helping you better understand that content. In this guide are practice test questions so that you can see how well you have grasped the content. Then, answer explanations are provided so that you can understand why you missed certain questions.

Don't try to cram the night before you take your exam. This is not a wise strategy for a few reasons. First, your retention of the information will be low. Your time would be better used by reviewing information you already know rather than trying to learn a lot of new information. Second, you will likely become stressed as you try to gain a large amount of knowledge in a short amount of time. Third, you will be depriving yourself of sleep. So be sure to go to bed at a reasonable time the night before. Being well-rested helps you focus and remain calm.

Be sure to eat a substantial breakfast the morning of the exam. If you are taking the exam in the afternoon, be sure to have a good lunch as well. Being hungry is distracting and can make it difficult to focus. You have hopefully spent lots of time preparing for the exam. Don't let an empty stomach get in the way of success!

When travelling to the testing center, leave earlier than needed. That way, you have a buffer in case you experience any delays. This will help you remain calm and will keep you from missing your appointment time at the testing center.

Be sure to pace yourself during the exam. Don't try to rush through the exam. There is no need to risk performing poorly on the exam just so you can leave the testing center early. Allow yourself to use all of the allotted time if needed.

Remain positive while taking the exam even if you feel like you are performing poorly. Thinking about the content you should have mastered will not help you perform better on the exam.

Once the exam is complete, take some time to relax. Even if you feel that you need to take the exam again, you will be well served by some down time before you begin studying again. It's often easier to convince yourself to study if you know that it will come with a reward!

Test-Taking Strategies

1. Predicting the Answer

When you feel confident in your preparation for a multiple-choice test, try predicting the answer before reading the answer choices. This is especially useful on questions that test objective factual knowledge. By predicting the answer before reading the available choices, you eliminate the possibility that you will be distracted or led astray by an incorrect answer choice. You will feel more confident in your selection if you read the question, predict the answer, and then find your prediction among the answer choices. After using this strategy, be sure to still read all of the answer choices carefully and completely. If you feel unprepared, you should not attempt to predict the answers. This would be a waste of time and an opportunity for your mind to wander in the wrong direction.

2. Reading the Whole Question

Too often, test takers scan a multiple-choice question, recognize a few familiar words, and immediately jump to the answer choices. Test authors are aware of this common impatience, and they will sometimes prey upon it. For instance, a test author might subtly turn the question into a negative, or he or she might redirect the focus of the question right at the end. The only way to avoid falling into these traps is to read the entirety of the question carefully before reading the answer choices.

3. Looking for Wrong Answers

Long and complicated multiple-choice questions can be intimidating. One way to simplify a difficult multiple-choice question is to eliminate all of the answer choices that are clearly wrong. In most sets of answers, there will be at least one selection that can be dismissed right away. If the test is administered on paper, the test taker could draw a line through it to indicate that it may be ignored; otherwise, the test taker will have to perform this operation mentally or on scratch paper. In either case, once the obviously incorrect answers have been eliminated, the remaining choices may be considered. Sometimes identifying the clearly wrong answers will give the test taker some information about the correct answer. For instance, if one of the remaining answer choices is a direct opposite of one of the eliminated answer choices, it may well be the correct answer. The opposite of obviously wrong is obviously right! Of course, this is not always the case. Some answers are obviously incorrect simply because they are irrelevant to the question being asked. Still, identifying and eliminating some incorrect answer choices is a good way to simplify a multiple-choice question.

4. Don't Overanalyze

Anxious test takers often overanalyze questions. When you are nervous, your brain will often run wild, causing you to make associations and discover clues that don't actually exist. If you feel that this may be a problem for you, do whatever you can to slow down during the test. Try taking a deep breath or counting to ten. As you read and consider the question, restrict yourself to the particular words used by the author. Avoid thought tangents about what the author *really* meant, or what he or she was *trying* to say. The only things that matter on a multiple-choice test are the words that are actually in the question. You must avoid reading too much into a multiple-choice question, or supposing that the writer meant something other than what he or she wrote.

5. No Need for Panic

It is wise to learn as many strategies as possible before taking a multiple-choice test, but it is likely that you will come across a few questions for which you simply don't know the answer. In this situation, avoid panicking. Because most multiple-choice tests include dozens of questions, the relative value of a single wrong answer is small. As much as possible, you should compartmentalize each question on a multiple-choice test. In other words, you should not allow your feelings about one question to affect your success on the others. When you find a question that you either don't understand or don't know how to answer, just take a deep breath and do your best. Read the entire question slowly and carefully. Try rephrasing the question a couple of different ways. Then, read all of the answer choices carefully. After eliminating obviously wrong answers, make a selection and move on to the next question.

6. Confusing Answer Choices

When working on a difficult multiple-choice question, there may be a tendency to focus on the answer choices that are the easiest to understand. Many people, whether consciously or not, gravitate to the answer choices that require the least concentration, knowledge, and memory. This is a mistake. When you come across an answer choice that is confusing, you should give it extra attention. A question might be confusing because you do not know the subject matter to which it refers. If this is the case, don't eliminate the answer before you have affirmatively settled on another. When you come across an answer choice of this type, set it aside as you look at the remaining choices. If you can confidently assert that one of the other choices is correct, you can leave the confusing answer aside. Otherwise, you will need to take a moment to try to better understand the confusing answer choice. Rephrasing is one way to tease out the sense of a confusing answer choice.

7. Your First Instinct

Many people struggle with multiple-choice tests because they overthink the questions. If you have studied sufficiently for the test, you should be prepared to trust your first instinct once you have carefully and completely read the question and all of the answer choices. There is a great deal of research suggesting that the mind can come to the correct conclusion very quickly once it has obtained all of the relevant information. At times, it may seem to you as if your intuition is working faster even than your reasoning mind. This may in fact be true. The knowledge you obtain while studying may be retrieved from your subconscious before you have a chance to work out the associations that support it. Verify your instinct by working out the reasons that it should be trusted.

8. Key Words

Many test takers struggle with multiple-choice questions because they have poor reading comprehension skills. Quickly reading and understanding a multiple-choice question requires a mixture of skill and experience. To help with this, try jotting down a few key words and phrases on a piece of scrap paper. Doing this concentrates the process of reading and forces the mind to weigh the relative importance of the question's parts. In selecting words and phrases to write down, the test taker thinks about the question more deeply and carefully. This is especially true for multiple-choice questions that are preceded by a long prompt.

9. Subtle Negatives

One of the oldest tricks in the multiple-choice test writer's book is to subtly reverse the meaning of a question with a word like *not* or *except*. If you are not paying attention to each word in the question, you can easily be led astray by this trick. For instance, a common question format is, "Which of the following is...?" Obviously, if the question instead is, "Which of the following is not...?," then the answer will be quite different. Even worse, the test makers are aware of the potential for this mistake and will include one answer choice that would be correct if the question were not negated or reversed. A test taker who misses the reversal will find what he or she believes to be a correct answer and will be so confident that he or she will fail to reread the question and discover the original error. The only way to avoid this is to practice a wide variety of multiple-choice questions and to pay close attention to each and every word.

10. Reading Every Answer Choice

It may seem obvious, but you should always read every one of the answer choices! Too many test takers fall into the habit of scanning the question and assuming that they understand the question because they recognize a few key words. From there, they pick the first answer choice that answers the question they believe they have read. Test takers who read all of the answer choices might discover that one of the latter answer choices is actually *more* correct. Moreover, reading all of the answer choices can remind you of facts related to the question that can help you arrive at the correct answer. Sometimes, a misstatement or incorrect detail in one of the latter answer choices will trigger your memory of the subject and will enable you to find the right answer. Failing to read all of the answer choices is like not reading all of the items on a restaurant menu: you might miss out on the perfect choice.

11. Spot the Hedges

One of the keys to success on multiple-choice tests is paying close attention to every word. This is never truer than with words like almost, most, some, and sometimes. These words are called "hedges" because they indicate that a statement is not totally true or not true in every place and time. An absolute statement will contain no hedges, but in many subjects, the answers are not always straightforward or absolute. There are always exceptions to the rules in these subjects. For this reason, you should favor those multiple-choice questions that contain hedging language. The presence of qualifying words indicates that the author is taking special care with his or her words, which is certainly important when composing the right answer. After all, there are many ways to be wrong, but there is only one way to be right! For this reason, it is wise to avoid answers that are absolute when taking a multiple-choice test. An absolute answer is one that says things are either all one way or all another. They often include words like *every*, *always*, *best*, and *never*. If you are taking a multiple-choice test in a subject that doesn't lend itself to absolute answers, be on your guard if you see any of these words.

12. Long Answers

In many subject areas, the answers are not simple. As already mentioned, the right answer often requires hedges. Another common feature of the answers to a complex or subjective question are qualifying clauses, which are groups of words that subtly modify the meaning of the sentence. If the question or answer choice describes a rule to which there are exceptions or the subject matter is complicated, ambiguous, or confusing, the correct answer will require many words in order to be expressed clearly and accurately. In essence, you should not be deterred by answer choices that seem excessively long. Oftentimes, the author of the text will not be able to write the correct answer without offering some qualifications and modifications. Your job is to read the answer choices thoroughly and completely and to select the one that most accurately and precisely answers the question.

13. Restating to Understand

Sometimes, a question on a multiple-choice test is difficult not because of what it asks but because of how it is written. If this is the case, restate the question or answer choice in different words. This process serves a couple of important purposes. First, it forces you to concentrate on the core of the question. In order to rephrase the question accurately, you have to understand it well. Rephrasing the question will concentrate your mind on the key words and ideas. Second, it will present the information to your mind in a fresh way. This process may trigger your memory and render some useful scrap of information picked up while studying.

14. True Statements

Sometimes an answer choice will be true in itself, but it does not answer the question. This is one of the main reasons why it is essential to read the question carefully and completely before proceeding to the answer choices. Too often, test takers skip ahead to the answer choices and look for true statements. Having found one of these, they are content to select it without reference to the question above. Obviously, this provides an easy way for test makers to play tricks. The savvy test taker will always read the entire question before turning to the answer choices. Then, having settled on a correct answer choice, he or she will refer to the original question and ensure that the selected answer is relevant. The mistake of choosing a correct-but-irrelevant answer choice is especially common on questions related to specific pieces of objective knowledge. A prepared test taker will have a wealth of factual knowledge at his or her disposal, and should not be careless in its application.

15. No Patterns

One of the more dangerous ideas that circulates about multiple-choice tests is that the correct answers tend to fall into patterns. These erroneous ideas range from a belief that B and C are the most common right answers, to the idea that an unprepared test-taker should answer "A-B-A-C-A-D-A-B-A." It cannot be emphasized enough that pattern-seeking of this type is exactly the WRONG way to approach a multiple-choice test. To begin with, it is highly unlikely that the test maker will plot the correct answers according to some predetermined pattern. The questions are scrambled and delivered in a random order. Furthermore, even if the test maker was following a pattern in the assignation of correct answers, there is no reason why the test taker would know which pattern he or she was using. Any attempt to discern a pattern in the answer choices is a waste of time and a distraction from the real work of taking the test. A test taker would be much better served by extra preparation before the test than by reliance on a pattern in the answers.

FREE DVD OFFER

Don't forget that doing well on your exam includes both understanding the test content and understanding how to use what you know to do well on the test. We offer a completely FREE Test Taking Tips DVD that covers world class test taking tips that you can use to be even more successful when you are taking your test.

All that we ask is that you email us your feedback about your study guide. To get your **FREE Test Taking Tips DVD**, email freedvd@studyguideteam.com with "FREE DVD" in the subject line and the following information in the body of the email:

- The title of your study guide.
- Your product rating on a scale of 1-5, with 5 being the highest rating.
- Your feedback about the study guide. What did you think of it?
- Your full name and shipping address to send your free DVD.

Overview of Exam

The concept of using a civil service exam for government employment dates back to the nineteenth century when anti-corruption reformers were seeking to eliminate the patronage system. Prior to the introduction of civil service exams, presidential administrations would use civil servant positions to reward loyalty. On January 16, 1883, Congress passed the Pendleton Civil Service Reform Act to establish the United States Civil Service Commission, which sought to create a merit system for federal employment by requiring applicants to pass a civil service exam. In the present day, although some federal agencies have decided to stop using the exams, many have continued the practice. For example, the United States Postal Service, Internal Revenue Service, and Federal Bureau of Investigation are among the agencies that require applicants to pass a civil service exam. The Office of Personnel Management currently oversees all federal employment practices, including the use of civil service exams. In addition, many state and local governments still use civil service exams to find qualified applicants.

The substance of civil service exams differs based on the specific federal agency, state government, or local government offering the exam. As such, there's no uniform civil service exam. For example, certain civil service exams might test for specialized skills, such as knowledge related to financial services or law enforcement. However, nearly all civil service exams have sections that cover the following topics: basic English language, problem-solving, interpersonal, and clerical skills.

Basic English language skills sections involve grammar, spelling, vocabulary, analogies, and reading comprehension. Depending on the exam, each topic might have an entire section devoted to it, or several of the topics could be combined. The grammar section could include verb tense, pronouns, and punctuation. The spelling section will ask which of several choices is the correct way to spell a word. Grammar and spelling are sometimes tested in a single section, such as proofreading a reading passage. The vocabulary section might simply ask test takers to match a word with its definition or ask what word best fits in context. The analogy section provides two words or a phrase in the prompt and then asks test takers to choose the answer mirroring that relationship. The reading comprehension section evaluates applicants' ability to process and understand written information. When answering reading comprehension questions, the answer is always based on the passage, not outside knowledge.

Problem-solving skills sections evaluate test takers' grasp on basic math skills and pattern recognition. Most math problems require test takers to complete simple equations or solve word problems by using addition, subtraction, multiplication, and division. Test takers will also need to have a strong handle on expressing numbers as fractions, percentages, and decimals. When a question asks for an "approximate" answer, it usually means the correct answer will be rounded to the nearest whole number. Other problem-solving sections feature a string of numbers and questions asking test takers to logically complete the pattern. Some civil service exams permit the use of a calculator, and if so, test takers will need to provide their own.

Interpersonal skills sections ask questions about hypothetical workplace situations. The questions will include the hypothetical in the prompt and then ask for the most appropriate and/or effective response. For positions where employees will have to engage with the public, this section is likely to include questions related to public education campaigns, public relations, and other forms of outreach to relevant stakeholders.

Clerical skills measure applicants' competency at basic office tasks, such as alphabetizing, completing forms, filing, following directions, reading charts and tables, and verifying names and numbers.

The overwhelming majority of civil service exam questions are multiple choice. The standard multiple-choice question includes a question in the prompt with four answer choices. Other types of multiple-choice questions will feature a word bank of answer choices. Less commonly, civil service exams might have a series of yes or no questions. For example, the spelling section might have questions that simply ask whether the word is spelled correctly. Civil service exams occasionally have a writing section, requiring a short or long written response, but those types of questions are typically for positions requiring specialized skills.

Prior to sitting for their civil service exam, test takers should know as much as possible about the specific exam's procedures and substantive content. They should know where the test is located, how travel should be arranged, and what materials they need to take to the test. Success requires avoiding these types of logistical failures, which can prevent applicants from sitting for the exam or undermine performance by adding unnecessary stress. Likewise, applicants should be similarly well versed on the substantive content, including the types of sections, question varieties, and time allotted for each section. Understanding what the exam emphasizes is the first step in creating a study plan.

Once applicants know how their exam weighs each section, the next step is to complete a set of practice questions for each section. Following this diagnostic test, test takers can create a study plan that focuses on their weaknesses, particularly in areas that are most likely to be tested. At the end of their preparation, test takers should complete sections under the actual time restraints to learn how long it takes to answer each type of question.

While taking the exam, test takers should follow a plan of attack. For example, if test takers struggle with a specific question type, they should set a time limit for those questions or save them for the end. Test takers should also follow the same strategy for filling out answer sheets, such as transferring the answer after every question or page in the exam booklet. On multiple-choice questions, test takers should use the process of elimination, crossing out clearly incorrect answer choices to increase their odds of selecting the correct answer. Lastly, test takers should make sure they answer every question. Most civil service exams don't penalize incorrect answers, so test takers should make an educated guess between the answer choices they haven't eliminated.

Spelling

Spelling might or might not be important to some, or maybe it just doesn't come naturally, but those who are willing to discover some new ideas and consider their benefits can learn to spell better and improve their writing. Misspellings reduce a writer's credibility and can create misunderstandings. Spell checkers built into word processors are not a substitute for accuracy. They are neither foolproof nor without error. In addition, a writer's misspelling of one word may also be a valid (but incorrect) word. For example, a writer intending to spell *herd* might accidentally type *s* instead of *d* and unintentionally spell *hers*. Since *hers* is a word, it would not be marked as a misspelling by a spell checker. In short, writers should use spell check, but not rely on it.

Guidelines for Spelling

Saying and listening to a word serves as the beginning of knowing how to spell it. Writers should keep these subsequent guidelines in mind, remembering there are often exceptions because the English language is replete with them.

Guideline #1: Syllables must have at least one vowel. In fact, every syllable in every English word has a vowel.

- d*o*g
- h*a*yst*a*ck
- *a*nsw*e*r*i*ng
- *a*bst*e*nt*iou*s
- s*i*mpl*e*

Guideline #2: The long and short of it. When the vowel has a short vowel sound as in *mad* or *bed,* only the single vowel is needed. If the word has a long vowel sound, add another vowel, either alongside it or separated by a consonant: bed/*bead*; mad/*made.* When the second vowel is separated by two spaces—*madder*—it does not affect the first vowel's sound.

Guideline #3: Suffixes. A **suffix** is a letter or group of letters added at the end of a word to form another word. The word created from the root and suffix is either a different tense of the same root (*help + ed = helped*) or a new word (*help + ful = helpful*). When written alone, suffixes are preceded by a dash to indicate that the root word comes before.

Some of the most common suffixes are the following:

Suffix	Meaning	Example
-ed	makes a verb past tense	washed
-ing	makes a verb a present participle verb	washing
-ly	to make characteristic of	lovely
-s, -es	to make more than one	chairs, boxes
-able	can be done	deplorable
-al	having characteristics of	comical
-est	comparative	greatest
-ful	full of	wonderful
-ism	belief in	communism
-less	Without	faithless
-ment	action or process	accomplishment
-ness	state of	happiness
-ize, -ise	to render, to make	sterilize, advertise
-cede, -ceed, -sede	Go	concede, proceed, supersede

Here are some helpful tips:

- When adding a suffix that starts with a vowel (for example, -ed) to a one-syllable root whose vowel has a short sound and ends in a consonant (for example, *stun*), the final consonant of the root (*n*) gets doubled.

    - stun + ed = stun*n*ed

- Exception: If the past tense verb ends in *x* such as *box*, the *x* does not get doubled.

    - box + ed = boxed

- If adding a suffix that starts with a vowel (-er) to a multi-syllable word ending in a consonant (*begin*), the consonant (*n*) is doubled.

    - begin + er = begin*n*er

- If a short vowel is followed by two or more consonants in a word such as *i+t+c+h = itch,* the last consonant should not be doubled.

    - itch + ed = itched

- If adding a suffix that starts with a vowel (-ing) to a word ending in *e* (for example, *name*), that word's final *e* is generally (but not always) dropped.

    - name + ing = naming
    - exception: manage + able = manageable

- If adding a suffix that starts with a consonant (-*ness*) to a word ending in *e* (*complete*), the *e* generally (but not always) remains.
  - complete + ness = completeness
  - exception: judge + ment = judgment
- There is great diversity on handling words that end in *y*. For words ending in a vowel + *y*, nothing changes in the original word.
  - play + ed = played
- For words ending in a consonant + *y*, the *y* id changed to i when adding any suffix except for –*ing*.
  - marry + ed = married
  - marry + ing = marrying

Guideline #4: Which comes first; the *i* or the *e*? Remember the saying, "*I* before e except after c or when sounding as *a* as in *neighbor* or *weigh*." Keep in mind that these are only guidelines and that there are always exceptions to every rule.

Guideline #5: Vowels in the right order. Another helpful rhyme is, "When two vowels go walking, the first one does the talking." When two vowels are in a row, the first one often has a long vowel sound and the other is silent. An example is *team*.

If one has difficulty spelling words, he or she can determine a strategy to help. Some people work on spelling by playing word games like Scrabble or Words with Friends®. Others use phonics, which is sounding words out by slowly and surely stating each syllable. People try repeating and memorizing spellings as well as picturing words in their head, or they may try making up silly memory aids. Each person should experiment and see what works best.

Homophones

Homophones are two or more words that have no particular relationship to one another except their identical pronunciations. Homophones make spelling English words fun and challenging. Examples include:

Common Homophones
affect, effect
allot, a lot
barbecue, barbeque
bite, byte
brake, break
capital, capitol
cash, cache
cell, sell
colonel, kernel
do, due, dew
dual, duel
eminent, imminent
flew, flu, flue
gauge, gage
holy, wholly
it's, its
knew, new
libel, liable
principal, principle
their, there, they're
to, too, two
yoke, yolk

Irregular Plurals

Irregular plurals are words that aren't made plural the usual way.

- Most nouns are made plural by adding –*s* (book*s*, television*s*, skyscraper*s*).

- Most nouns ending in *ch, sh, s, x,* or *z* are made plural by adding –*es* (church*es*, marsh*es*).

- Most nouns ending in a vowel + *y* are made plural by adding –*s* (day*s*, toy*s*).

- Most nouns ending in a consonant + *y,* are made plural by the -*y* becoming -*ies* (baby becomes *babies*).

- Most nouns ending in an *o* are made plural by adding –*s* (piano*s*, photo*s*).

- Some nouns ending in an *o*, though, may be made plural by adding –*es* (example: potato*es*, volcano*es*), and, of note, there is no known rhyme or reason for this!

- Most nouns ending in an *f* or *fe* are made plural by the *-f* or *-fe* becoming *-ves*! (example: wolf becomes *wolves*).

- Some words function as both the singular and plural form of the word (fish, deer).

- Other exceptions include *man* becomes *men, mouse* becomes *mice, goose* becomes *geese,* and *foot* becomes *feet.*

Contractions

The basic rule for making **contractions** is one area of spelling that is pretty straightforward: combine the two words by inserting an apostrophe (') in the space where a letter is omitted. For example, to combine *you* and *are*, drop the *a* and put the apostrophe in its place: *you're.*

he + is = he's

you + all = y'all (informal, but often misspelled)

Note that *it's*, when spelled with an apostrophe, is always the contraction for *it is*. The possessive form of the word is written without an apostrophe as *its*.

Correcting Misspelled Words

A good place to start looking at commonly misspelled words here is with the word *misspelled*. While it looks peculiar, look at it this way: *mis* (the prefix meaning *wrongly*) + *spelled* = *misspelled*.

Let's look at some commonly misspelled words.

Commonly Misspelled Words					
accept	benign	existence	jewelry	parallel	separate
acceptable	bicycle	experience	judgment	pastime	sergeant
accidentally	brief	extraordinary	library	permissible	similar
accommodate	business	familiar	license	perseverance	supersede
accompany	calendar	February	maintenance	personnel	surprise
acknowledgement	campaign	fiery	maneuver	persuade	symmetry
acquaintance	candidate	finally	mathematics	possess	temperature
acquire	category	forehead	mattress	precede	tragedy
address	cemetery	foreign	millennium	prevalent	transferred
aesthetic	changeable	foremost	miniature	privilege	truly
aisle	committee	forfeit	mischievous	pronunciation	usage
altogether	conceive	glamorous	misspell	protein	valuable
amateur	congratulations	government	mortgage	publicly	vengeance
apparent	courtesy	grateful	necessary	questionnaire	villain
appropriate	deceive	handkerchief	neither	recede	Wednesday
arctic	desperate	harass	nickel	receive	weird
asphalt	discipline	hygiene	niece	recommend	
associate	disappoint	hypocrisy	ninety	referral	
attendance	dissatisfied	ignorance	noticeable	relevant	
auxiliary	eligible	incredible	obedience	restaurant	
available	embarrass	intelligence	occasion	rhetoric	
balloon	especially	intercede	occurrence	rhythm	
believe	exaggerate	interest	omitted	schedule	
beneficial	exceed	irresistible	operate	sentence	

Practice Questions

For each question below, pick out the word that is spelled incorrectly:

1. a. alliance   b. interpret   c. militery   d. cylinder

2. a. legitimate   b. mecanical   c. statue   d. restaurant

3. a. subsequent   b. tenent   c. naïve   d. checked

4. a. illegal   b. opponent   c. handkerchief   d. dissapproval

5. a. ignore   b. eligible   c. ladder   d. citisen

6. a. empty   b. adequate   c. cardbord   d. nocturnal

7. a. inferior   b. individule   c. normally   d. inconsistent

8. a. gorgeous   b. transfer   c. momentery   d. recommendation

9. a. preliminary   b. natural   c. worthie   d. pencils

10. a. anxious   b. vacation   c. eruption   d. spekulate

11. a. devision   b. interim   c. metropolitan   d. console

12. a. laundry   b. plead   c. qwash   d. streaked

13. a. ocurred   b. patrol   c. actual   d. detrimental

14. a. enjoyable   b. frightened   c. perspiration   d. unbarable

15. a. thirsty   b. warehouse   c. promis   d. fortunately

16. a. bandage   b. siezed   c. difference   d. counselor

17. a. draught   b. despicable   c. enormous   d. answerd

18. a. expenses   b. calender   c. disguise   d. procrastination

19. a. holistic   b. candidate   c. known   d. adolesent

20. a. massacre   b. fatigue   c. detaled   d. congratulate

21. a. fixture   b. health   c. jurer   d. blemish

22. a. gradually   b. inderect   c. equation   d. vacillate

23. a. optimistic   b. literally   c. managment   d. amused

24. a. natural   b. obviously   c. perseption   d. miniscule

25. a. curiosity   b. quell   c. release   d. pasify

26. a. syringe   b. pivet   c. subject   d. potential

27. a. suggest   b. withdrawl   c. reduction   d. tension

28. a. mobile   b. ought   c. reality   d. hardwear

29. a. mold   b. instrament   c. offender   d. panic

30. a. palm   b. official   c. miniture   d. contemplate

31. a. kneel   b. introduction   c. novealty   d. familiar

32. a. railing   b. modern   c. oponent   d. prejudice

33. a. barrister   b. repitition   c. sign   d. optimal

34. a. comittment   b. align   c. guest   d. intrude

35. a. fierce   b. haste   c. elimanate   d. approach

36. a. amatur   b. campaign   c. bureau   d. scalding

37. a. mediation   b. comparible   c. niche   d. apologize

38. a. column   b. narow   c. nurture   d. whisper

39. a. effective   b. fluorescent   c. capital   d. garantee

40. a. entergetic   b. flowing   c. smear   d. leader

41. a. depth   b. subsequent   c. soliciter   d. decorative

42. a. junior   b. intrusion   c. leverage   d. nausious

43. a. pause   b. ilustrate   c. minister   d. existence

44. a. murmer   b. staple   c. typical   d. traveler

45. a. secreteries   b. accidentally   c. chaos   d. legendary

46. a. height   b. industrial   c. lining   d. gorgious

47. a. paragraph   b. offender   c. mayer   d. versatile

48. a. contraversy   b. monument   c. keyboard   d. overbearing

49. a. recreation   b. collision   c. element   d. superviser

50. a. service   b. merit   c. insurance   d. violense

Answer Explanations

1. C: The word *militery* should be spelled *military*. Alliance, interpret, and cylinder are the correct spellings.

2. B: The word *mecanical* should be spelled *mechanical*. Legitimate, statue, and restaurant are spelled correctly.

3. B: The word *tenent* should be *tenant*. Subsequent, naïve, and checked are spelled correctly.

4. D: The word *dissapproval* should be *disapproval*. Illegal, opponent, and handkerchief are the correct spellings.

5. D: The word *citisen* should be *citizen*. Ignore, eligible, and ladder are spelled correctly.

6. C: The word *cardbord* should be *cardboard*. Empty, adequate, and nocturnal are spelled correctly.

7. B: The word *individule* should be individual. Inferior, normally, and inconsistent are the correct spellings.

8. C: The word *momentery* should be *momentary*. Gorgeous, transfer, and recommendation are spelled correctly.

9. C: The word *worthie* should be spelled *worthy*. Preliminary, natural, and pencils are spelled correctly.

10. D: The word *spekulate* should be spelled *speculate*. Anxious, vacation, and eruption are spelled correctly.

11. A: The word *devision* should be *division*. Interim, metropolitan, and console are spelled correctly.

12. C: The word *qwash* should be *quash*. Laundry, plead, and streaked are spelled correctly.

13. A: The word *ocurred* should be occurred. Patrol, actual, and detrimental are the correct spellings.

14. D: The word *unbarable* should be *unbearable*. Enjoyable, frightened, and perspiration are spelled correctly.

15. C: The word *promis* should be *promise*. Thirsty, warehouse, and promise are spelled correctly.

16. B: The word *siezed* should be *seized*. Remember, *I* before *e* except after *c*. Bandage, difference, and counselor are spelled correctly.

17. D: The word *answerd* should be *answered*. Draught , despicable, and enormous are spelled correctly.

18. B: The word *calender* should be *calendar*. Expenses, disguise, and procrastination are spelled correctly.

19. D: The word *adolesent* should be *adolescent*. Holistic, candidate, and known are spelled correctly.

20. C: The word *detaled* should be *detailed*. Massacre, fatigue, and congratulate are spelled correctly.

21. C: The word *jurer* should be *juror*. Fixture, health, and blemish are spelled correctly.

22. B: The word *inderect* should be *indirect*. Gradually, equation, and vacillate are the correct spellings.

23. C: The word *managment* should be *management*. Optimistic, literally, and amused are spelled correctly.

24. C: The word *perseption* should be *perception*. Natural, obviously, and miniscule are the correct spellings

25. D: The word *pasify* should be *pacify*. Curiosity, quell, and release are spelled correctly.

26. B: The word *pivet* should be *pivot*. Syringe, subject, and potential are spelled correctly.

27. B: The word *withdrawl* should be *withdrawal*. Suggest, reduction, and tension are the correct spellings.

28. D: The word *hardwear* should be *hardware*. Mobile, ought, and reality are spelled correctly.

29. B: The word *instrament* should be *instrument*. Mold, offender, and panic are spelled correctly.

30. C: The word *miniture* should be *miniature*. Palm, official, and contemplate are spelled correctly.

31. C: The word *novealty* should be *novelty*. Kneel, introduction, and familiar are the correct spellings.

32. C: The word *oponent* should be *opponent*. Railing, modern, and prejudice are the correct spellings.

33. B: The word *repitition* should be repetition. Barrister, sign, and optimal are spelled correctly.

34. A: The word *comittment* should be *commitment*. Align, guest, and intrude are spelled correctly.

35. C: The word *elimanate* should be *eliminate*. Fierce, hate, and approach are the correct spellings.

36. A: The word *amatur* should be *amateur*. Campaign, bureau, and scalding are correct spellings.

37. B: The word *comparible* should be *comparable*. Mediation, niche, and apologize are correct spellings.

38. B: The word *narow* should be *narrow*. Column, nurture, and whisper are spelled correctly.

39. D: The word *garantee* should be *guarantee*. Effective, fluorescent, and capital are spelled correctly.

40. A: The word *entergetic* should be *energetic*. Flowing, smear, and leader are spelled correctly.

41. C: The word *soliciter* should be *solicitor*. Depth, subsequent, and decorative are spelled correctly.

42. D: The word *nausious* should be *nauseous*. Junior, intrusion, and leverage are spelled correctly.

43. B: The word *ilustrate* should be *illustrate*. Pause, minister, and existence are spelled correctly.

44. A: The word *murmer* should be *murmur*. Staple, typical, and traveler are spelled correctly.

45. A: The word *secreteries* should be *secretaries*. Accidentally, chaos, and legendary are spelled correctly.

46. D: The word *gorgious* should be *gorgeous*. Height, industrial, and lining are spelled correctly.

47. C: The word *mayer* should be *mayor*. Paragraph, offender, and versatile are spelled correctly.

48. A: The word *contraversy* should be *controversy*. Monument, keyboard, and overbearing are spelled correctly.

49. D: The word *superviser* should be *supervisor*. Recreation, collision, and element are spelled correctly.

50. D: The word *violense* should be *violence*. Service, merit, and insurance are spelled correctly.

Vocabulary

Parts of Speech

Also referred to as **word classes**, **parts of speech** refer to the various categories in which words are placed. Words can be placed in any one or a combination of the following categories:

- Nouns
- Determiners
- Pronouns
- Verbs
- Adjectives
- Adverbs
- Prepositions
- Conjunctions

Understanding the various parts of speech used in the English language helps readers to better understand the written language.

<u>Nouns</u>
Nouns are defined as any word that represents a person, place, animal, object, or idea. Nouns can identify a person's title or name, a person's gender, and a person's nationality, such as *banker, commander-in-chief, female, male,* or *an American*.

With animals, nouns identify the kingdom, phylum, class, etc. For example: animal is *elephant*; phylum is *Chordata*; or class is *Mammalia*.

When identifying places, nouns refer to a physical location, a general vicinity, or the proper name of a city, state, or country. Some examples include the *desert, the East, Phoenix, Arizona,* or *the United States*.

There are eight types of nouns: common, proper, countable, uncountable, concrete, abstract, compound, and collective.

Common nouns are used in general terms, without specific identification. Examples include *girl, boy, country,* or *school*. **Proper nouns** refer to the proper name given to people, places, animals, or entities, such as *Melissa, Martin, Italy,* or *Harvard*.

Countable nouns can be counted: one car, two cars, or three cars. **Uncountable nouns** cannot be counted, such as *air, liquid,* or *gas*.

To be abstract is to exist, but only in thought or as an idea. An **abstract noun** cannot be physically touched, seen, smelled, heard, or tasted. These include *chivalry, day, fear, thought, truth, friendship,* or *freedom*.

To be concrete is to be seen, touched, tasted, heard, and/or smelled. Examples of **concrete nouns** include *pie, snow, tree, bird, desk, hair,* or *dog*.

A **compound noun** is another term for an open compound word. Any noun that is written as two nouns that together form a specific meaning is a compound noun, such as *post office, ice cream,* or *swimming pool.*

A **collective noun** is formed by grouping three or more words to create one meaning. Examples include *bunch of flowers, herd of elephants, flock of birds,* or *school of fish.*

Determiners

Determiners modify a noun and usually refer to something specific. Determiners fall into one of four categories: articles, demonstratives, quantifiers, or possessive determiners.

Articles can be both definite articles, as in *the*, and indefinite as in *a, an, some,* or *any*:

> *The* man with *the* red hat.

> *A* flower growing in *the* yard.

> *Any* person who visits *the* store.

There are four different types of **demonstratives**: *this, that, these,* and *those.*

True demonstrative words will not directly precede the noun of the sentence, but will *be* the noun. Some examples:

> *This* is the one.

> *That* is the place.

> *Those* are the files.

Once a demonstrative is placed directly in front of the noun, it becomes a **demonstrative pronoun**:

> *This* one is perfect.

> *That* place is lovely.

> *Those* boys are annoying.

Quantifiers proceed nouns to give additional information for how much or how many. They can be used with countable and uncountable nouns:

> She bought *plenty* of apples.

> *Few* visitors came.

> I got a *little* change.

Possessive determiners, sometimes called **possessive adjectives**, indicate possession. They are the possessive forms of personal pronouns, such as *my, your, hers, his, its, their,* or *our*:

> That is *my* car.

> Tom rode *his* bike today.

> Those papers are *hers.*

Pronouns

Pronouns are words that stand in place of nouns. There are three different types of pronouns: **subjective pronouns** (*I, you, he, she, it, we, they*), **objective pronouns** (*me, you, him, her it, us, them*), and **possessive pronouns** (*mine, yours, his, hers, ours, theirs*).

You'll see some words are found in more than one pronoun category. See examples and clarifications below:

> *You* are going to the movies.

You is a subjective pronoun; it is the subject of the sentence and is performing the action.

> I threw the ball to *you.*

You is an objective pronoun; it is receiving the action and is the object of the sentence.

> We saw *her* at the movies.

Her is an objective pronoun; it is receiving the action and is the object of the sentence.

> The house across the street from the park is *hers.*

Hers is a possessive pronoun; it shows possession of the house and is used as a possessive pronoun.

Verbs

Verbs are words in a sentence that show action or state. Just as there can be no sentence without a subject, there can be no sentence without a verb. In these sentences, notice verbs in the present, past, and future tenses. To form some tenses in the future and in the past requires auxiliary, or helping, verbs:

> I *see* the neighbors across the street.

See is an action.

> We *were eating* at the picnic.

Eating is the main action, and the verb *were* is the past tense of the verb *to be*, and is the helping or auxiliary verb that places the sentence in the past tense.

> You *will turn* 20 years of age next month.

Turn is the main verb, but to show a future tense, *will* is the helping verb to show future tense of the verb *to be*.

Adjectives

Adjectives are a special group of words used to modify or describe a noun. Adjectives provide more information about the noun they modify. For example:

> The boy went to school. (There is no adjective.)

Rewrite the sentence, adding an adjective to further describe the boy and/or the school:

> The *young* boy went to the *old* school. (The adjective *young* describes the boy, and the adjective *old* describes the school.)

Adverbs

Adverb can play one of two roles: to modify the adjective or to modify the verb. For example:

> The young boy went to the old school.

We can further describe the adjectives *young* and *old* with adverbs placed directly in front of the adjectives:

> The *very* young boy went to the *very* old school. (The adverb *very* further describes the adjectives *young* and *old*.)

Other examples of using adverbs to further describe verbs:

> The boy *slowly* went to school.

> The boy *quickly* went to school.

The adverbs *slowly* and *quickly* further modify the verbs.

Prepositions

Prepositions are special words that generally precede a noun. Prepositions clarify the relationship between the subject and another word or element in the sentence. They clarify time, place, and the positioning of subjects and objects in a sentence. Common prepositions in the English language include: *near, far, under, over, on, in, between, beside, of, at, until, behind, across, after, before, for, from, to, by,* and *with*.

Conjunctions

Conjunctions are a group of unique words that connect clauses or sentences. They also work to coordinate words in the same clause. It is important to choose an appropriate conjunction based on the meaning of the sentence. Consider these sentences:

> I really like the flowers, *however* the smell is atrocious.

> I really like the flowers, *besides* the smell is atrocious.

The conjunctions *however* and *besides* act as conjunctions, connecting the two ideas: *I really like the flowers,* and, *the smell is atrocious.* In the second sentence, the conjunction *besides* makes no sense and would confuse the reader. You must consider the message you wish to convey and choose conjunctions that clearly state that message without ambiguity.

Some conjunctions introduce an opposing opinion, thought, or fact. They can also reinforce an opinion, introduce an explanation, reinforce cause and effect, or indicate time. For example:

She wishes to go to the movies, *but* she doesn't have the money. (Opposition)

The professor became ill, *so* the class was postponed. (Cause and effect)

They visited Europe *before* winter came. (Time)

Each conjunction serves a specific purpose in uniting two separate ideas. Below are common conjunctions in the English language:

Opposition	Cause & Effect	Reinforcement	Time	Explanation
however	therefore	besides	afterward	for example
nevertheless	as a result	anyway	before	in other words
but	because of this	after all	firstly	for instance
although	consequently	furthermore	next	such as

Practice Questions

Find the synonym, or the word closest in meaning, for the words represented in all caps below.

1. DEDUCE
   a. Explain
   b. Win
   c. Reason
   d. Gamble
   e. Undo

2. ELUCIDATE
   a. Learn
   b. Enlighten
   c. Plan
   d. Corroborate
   e. Conscious

3. VERIFY
   a. Criticize
   b. Change
   c. Teach
   d. Substantiate
   e. Resolve

4. INSPIRE
   a. Motivate
   b. Impale
   c. Exercise
   d. Patronize
   e. Collaborate

5. PERCEIVE
   a. Sustain
   b. Collect
   c. Prove
   d. Lead
   e. Comprehend

6. NOMAD
   a. Munching
   b. Propose
   c. Wanderer
   d. Conscientious
   e. Blissful

7. PERPLEXED
   a. Annoyed
   b. Vengeful
   c. Injured
   d. Confused
   e. Prepared

8. LYRICAL
   a. Whimsical
   b. Vague
   c. Fruitful
   d. Expressive
   e. Playful

9. BREVITY
   a. Dullness
   b. Dangerous
   c. Brief
   d. Ancient
   e. Calamity

10. IRATE
    a. Angry
    b. Knowledgeable
    c. Tired
    d. Confused
    e. Taciturn

11. LUXURIOUS
    a. Faded
    b. Bright
    c. Lavish
    d. Inconsiderate
    e. Overwhelming

12. IMMOBILE
    a. Fast
    b. Slow
    c. Eloquent
    d. Vivacious
    e. Sedentary

13. OVERBEARING
    a. Neglect
    b. Overactive
    c. Clandestine
    d. Formidable
    e. Amicable

14. PREVENT
    a. Avert
    b. Rejoice
    c. Endow
    d. Fulfill
    e. Ensure

15. REPLENISH
    a. Falsify
    b. Hindsight
    c. Dwell
    d. Refresh
    e. Nominate

16. REGALE
    a. Remember
    b. Grow
    c. Outnumber
    d. Entertain
    e. Bore

17. WEARY
    a. Tired
    b. Clothing
    c. Happy
    d. Hot
    e. Whiny

18. VAST
    a. Rapid
    b. Expansive
    c. Small
    d. Ocean
    e. Uniform

19. DEMONSTRATE
    a. Tell
    b. Show
    c. Build
    d. Complete
    e. Make

20. ORCHARD
    a. Flower
    b. Fruit
    c. Grove
    d. Peach
    e. Farm

21. TEXTILE
    a. Fabric
    b. Document
    c. Mural
    d. Ornament
    e. Knit

22. OFFSPRING
    a. Bounce
    b. Parent
    c. Music
    d. Child
    e. Skip

23. PERMIT
    a. Law
    b. Parking
    c. Crab
    d. Jail
    e. Allow

24. WOMAN
    a. Man
    b. Lady
    c. Women
    d. Girl
    e. Mother

25. ROTATION
    a. Wheel
    b. Year
    c. Spin
    d. Flip
    e. Orbit

26. CONSISTENT
    a. Stubborn
    b. Contains
    c. Sticky
    d. Texture
    e. Steady

27. PRINCIPLE
    a. Principal
    b. Leader
    c. President
    d. Foundation
    e. Royal

28. PERIMETER
   a. Outline
   b. Area
   c. Side
   d. Volume
   e. Inside

29. SYMBOL
   a. Drum
   b. Music
   c. Clang
   d. Emblem
   e. Text

30. GERMINATE
   a. Doctor
   b. Sick
   c. Infect
   d. Plants
   e. Grow

Answer Explanations

1. C: To deduce something is to figure it out using reasoning. Although this might cause a win and prompt an explanation to further understanding, the art of deduction is logical reasoning.

2. B: To elucidate, a light is figuratively shined on a previously unknown or confusing subject. This Latin root, "lux" meaning "light," prominently figures into the solution. *Enlighten* means to educate, or bring into the light.

3. D: Looking at the Latin word "veritas," meaning "truth," will yield a clue as to the meaning of *verify*. To verify is the act of finding or assessing the truthfulness of something. This usually means amassing evidence to substantiate a claim. *Substantiate*, of course, means to provide evidence to prove a point.

4. A: If someone is inspired, they are motivated to do something. Someone who is an inspiration motivates others to follow his or her example.

5. E: All the connotations of *perceive* involve the concept of seeing. Whether figuratively or literally, perceiving implies the act of understanding what is presented. The word *comprehend* is synonymous with this understanding.

6. C: Nomadic tribes are those who, throughout history and even today, prefer to wander their lands instead of settling in any specific place. *Wanderer* best describes these people.

7. D: *Perplexed* means baffled or puzzled, which are synonymous with *confused*.

8. D: *Lyrical* is used to refer to something being poetic or song-like, characterized by showing enormous imagination and description. While the context of lyrical can be playful or even whimsical, the best choice is *expressive*, since whatever emotion lyrical may be used to convey in context will be expressive in nature.

9. C: *Brevity* literally means brief or concise. Note the similar beginnings of *brevity* and *brief*—from the Latin "brevis," meaning brief.

10. A: *Irate* means being in a state of anger. Clearly this is a negative word that can be paired with another word in kind. The closest word to match this is obviously *angry*. Research would also reveal that irate comes from the Latin "ira," which means anger.

11. C: *Lavish* is a synonym for *luxurious*—both describe elaborate and/or elegant lifestyles and/or settings.

12. E: *Immobile* means "not able to move." The two best selections are *B* and *E*—but *slow* still implies some form of motion, whereas *sedentary* has the connotation of being seated and/or inactive for a significant portion of time and/or as a natural tendency.

13. B: *Overbearing* refers to domineering or being oppressive. This is emphasized in the "over" prefix, which emphasizes an excess in definitions. This prefix is also seen in *overactive*. Similar to overbearing, overacting reflects an excess of action.

14. A: The "pre" prefix describes something that occurs before an event. *Prevent* means to stop something before it happens. This leads to a word that relates to something occurring beforehand and, in a way, is preventive—wanting to stop something. *Avert* literally means to turn away or ward off an impending circumstance, making it the best fit.

15. D: *Refresh* is synonymous with *replenish*. Both words mean to restore or refill. Additionally, these terms do share the "re" prefix as well.

16. D: *Regale* literally means to amuse someone with a story. This is a very positive word; the best way to eliminate choices is to look for a term that matches regale in both positive context/sound and definition. *Entertain* is both a positive word and a synonym of regale.

17. A: *Weary* most closely means *tired*. Someone who is weary and tired may be whiny, but they do not necessarily mean the same thing.

18. B: Something that is *vast* is far-reaching and *expansive*. Choice *D, ocean*, may be described as vast but the word alone doesn't mean vast. The heavens or skies may also be described as vast. Someone's imagination or vocabulary can also be vast.

19. B: To demonstrate something means to show it. The word *demonstration* comes from demonstrate and a demonstration is a modeling or show-and-tell type of example that is usually visual.

20. C: An *orchard* is most like a *grove* because both are areas like plantations that grow different kinds of fruit. *Peach* is a type of fruit that may be grown in an orchard but it is not a synonym for *orchard*. Many citrus fruits are grown in groves but either word can be used to describe many fruit-bearing trees in one area. Choice *E, farm*, may have an orchard or grove on the property, but they are not the same thing and many farms do not grow fruit trees.

21. A: A *textile* is another word for a fabric. The most confusing alternative choice in this case is *knit*, because some textiles are knit but *textile* and *knit* are not synonyms and plenty of textiles are not knit.

22. D: *Offspring* are the children of parents. This word is especially common when talking about the animal kingdom, although it can be used with humans as well. *Offspring* does have the word *spring* in it, although it has nothing to do with bouncing or jumping. The other answer choice, *parent*, may be somewhat tricky because parents have offspring, but for this reason, they are not synonyms.

23. E: *Permit* can be a verb or a noun. As a verb, it means to allow or give authorization for something. As a noun, it generally refers to a document or something that has been authorized like a parking permit or driving permit, allowing the authorized individual to park or drive under the rules of the document.

24. B: A *woman* is a lady. Test takers must read carefully and remember the difference between *woman* and *women*. *Woman* refers to an individual lady or one person who is female, while *women* is the plural form and refers to more than one, or a group, of ladies. A woman may be a mother but not necessarily, and these words are not synonyms. A girl is a child and not yet a woman.

25. C: *Rotation* means to spin or turn, such as a wheel rotating on a car, although wheel does not *mean* rotation.

26. E: Something that is consistent is steady, predictable, reliable, or constant. The tricky one here is that the word *consistency* comes from the word consistent, and may describe a text or something that is sticky. *Consistent* also comes from the word *consist*, which means to contain (Choice *B*). Test takers must be discerning readers and knowledgeable about vocabulary to recognize the difference in these words and look for the true synonym of *consistent*.

27. D: A *principle* is a foundation or a guiding idea or belief. Someone with good moral character is described as having strong principles. Test takers must be careful not to get confused with the homonyms *principle* and *principal*, because these words have very different meanings. A *principal* is the leader of a school and the word *principal* also refers to the main or most important idea or thing.

28. A: *Perimeter* refers to the outline or borders of an object. Test takers may recognize that word from math class, where perimeter refers to the edges or distance around an enclosed shape. Some of the other answer choices refer to other math vocabulary encountered in geometry lessons, but do not have the same meaning as *perimeter.*

29. D: A *symbol* is an object, picture, or sign that is used to represent something. For example, a pink ribbon is a symbol for breast-cancer awareness and a flag can be a symbol for a country. The tricky part of this question was also knowing the meaning of *emblem,* which typically describes a design that represents a group or concept, much like a symbol. Emblems often appear on flags or a coat of arms.

30. E: *Germinate* means to develop or grow and most often refers to sprouting seeds as a new plant first breaks through the seed coat. It can also refer to the development of an idea. Choice *D* may be an attractive choice since plants germinate but germinate does not mean *plant.*

Analogies

What are Verbal Analogies?

Analogies compare two different things that have a relationship or some similarity. For example, a basic analogy is *apple is to fruit as cucumber is to vegetable*. This analogy points out the category under which each item falls. On most iterations of the Civil Service exam, the final term (*vegetable*, in this case) will be blank and must be filled in from the multiple choices, selecting the word that best demonstrates the relationship in the first pair of words. Other analogies include words that are **synonyms**, which are words that share similar meanings to one another. For example, *big* and *large* are synonyms and *tired* and *sleepy* are also synonyms. Verbal analogy questions can be difficult because they require the test taker to demonstrate an understanding of small differences and similarities in both word meanings and word relationships.

Layout of the Questions

Verbal analogy sections are on other standardized tests such as the SAT. The format on most tests is basically the same. First, two words are paired together that provide a frame for the analogy. Then, you are given a third word. You must find the relationship between the first two words, and then choose the fourth word to match the relationship to the third word. It may help to think of it like this: A is to B as C is to D. Examine the breakdown below:

Apple (A) is to fruit (B) as carrot (C) is to vegetable (D).

As shown above, there are four words: the first three are given and the fourth word is the answer that must be found. The first two words are given to set up the kind of analogy that is to be replicated for the next pair. We see that apple is paired with fruit. In the first pair, a specific food item, apple, is paired to the food group category it corresponds with, which is fruit. When presented with the third word in the verbal analogy, carrot, a word must be found that best matches carrot in the way that fruit matched with apple. Again, carrot is a specific food item, so a match should be found with the appropriate food group: vegetable! Here's a sample prompt:

Morbid is to dead as jovial is to

    a. Hate.
    b. Fear.
    c. Disgust.
    d. Happiness.
    e. Desperation.

As with the apple and carrot example, here is an analogy frame in the first two words: *morbid* and *dead*. Again, this will dictate how the next two words will correlate with one another. The definition of *morbid* is: described as or appealing to an abnormal and unhealthy interest in disturbing and unpleasant subjects, particularly death and disease. In other words, *morbid* can mean ghastly or death-like, which is why the word *dead* is paired with it. Dead relates to morbid because it describes morbid. With this in mind, *jovial* becomes the focus. Jovial means joyful, so out of all the choices given, the closest answer describing jovial is *happiness (Choice D)*.

Prompts on the exam will be structured just like the one above. "A is to B as C is to ?" will be given, where the answer completes the second pair. Or sometimes, "A is to B as ? is to ?" is given, where the second pair of words must be found that replicate the relationship between the first pair. The only things that will change are the words and the relationships between the words provided.

Discerning the Correct Answer

While it wouldn't hurt in test preparation for test takers to expand their vocabulary, verbal analogies are all about delving into the words themselves and finding the right connection—the right word that will fit an analogy. People preparing for the test shouldn't think of themselves as human dictionaries, but rather as detectives. Remember that the first two words that are connected dictates the second pair. From there, picking the correct answer or simply eliminating the ones that aren't correct is the best strategy.

Just like a detective, a test taker needs to carefully examine the first two words of the analogy for clues. It's good to get in the habit of asking the questions: What do the two words have in common? What makes them related or unrelated? How can a similar relationship be replicated with the word I'm given and the answer choices? Here's another example:

Pillage is to steal as meander is to

a. Stroll.
b. Burgle.
c. Cascade.
d. Accelerate.
e. Pinnacle.

Why is *pillage* paired with *steal*? In this example, pillage and steal are synonymous: they both refer to the act of stealing. This means that the answer is a word that means the same as *meander*, which is *stroll*. In this case, the defining relationship in the whole analogy was a similar definition.

What if test takers don't know what stroll or meander mean, though? Using logic helps to eliminate choices and pick the correct answer. Looking closer into the contexts of the words *pillage* and *steal,* here are a few facts: these are things that humans do; and while they are actions, these are not necessarily types of movement. Again, pick a word that will not only match the given word, but best completes the relationship. It wouldn't make sense that *burgle* (*B*) would be the correct choice because *meander* doesn't have anything to do with stealing, so that eliminates *burgle*. *Pinnacle* (*E*) also can be eliminated because this is not an action at all but a position or point of reference. *Cascade* (*C*) refers to pouring or falling, usually in the context of a waterfall and not in reference to people, which means we can eliminate cascade as well. While people do accelerate when they move, they usually do so under additional circumstances: they accelerate while running or driving a car. All three of the words we see in the analogy are actions that can be done independently of other factors. Therefore, *accelerate* (*D*) can be eliminated, and *stroll* (*A*) should be chosen. *Stroll* and *meander* both refer to walking or wandering, so this fits perfectly.

The process of elimination will help rule out wrong answers. However, the best way to find the correct answer is simply to differentiate the correct answer from the other choices. For this, test takers should go back to asking questions, starting with the chief question: What's the connection? There are actually many ways that connections can be found between words. The trick is to look for the answer that is consistent with the relationship between the words given. What is the prevailing connection? Here are a few different ways verbal analogies can be formed.

Finding Connections in Word Analogies

Categories

One of the easiest ways to choose the correct answer in word analogies is simply to group the words. Ask if the words can be compartmentalized into distinct categories. Here are some examples:

Terrier is to dog as mystery is to

    a. Thriller.
    b. Murder.
    c. Detective.
    d. Novel.
    e. Investigation.

This one might have been a little confusing, but when looking at the first two words in the analogy, this is clearly one in which a category is the prevailing theme. Think about it: a terrier is a type of dog. While there are several breeds of dogs that can be categorized as a terrier, in the end, all terriers are still dogs. Therefore, *mystery* needs to be grouped into a category. Murders, detectives, and investigations can all be involved in a mystery plot, but a *murder* (*B*), a *detective* (*C*), or an *investigation* (*E*) is not necessarily a mystery. A *thriller* (*A*) is a purely fictional concept, a kind of story or film, just like a mystery. A thriller can describe a mystery, but the same issue appears as the other choices. What about *novel* (*D*)? For one thing, it's distinct from all the other terms. A novel isn't a component of a mystery, but a mystery can be a type of novel. The relationship fits: a terrier is a type of dog, just like a mystery is a type of novel.

Synonym/Antonym

Some analogies are based on words meaning the same thing or expressing the same idea. Sometimes it's the complete opposite!

Marauder is to brigand as

    a. King is to peasant.
    b. Juice is to orange.
    c. Soldier is to warrior.
    d. Engine is to engineer.
    e. Paper is to photocopier.

Here, soldier is to warrior (*C*) is the correct answer. *Marauders* and *brigands* are both thieves. They are synonyms. The only pair of words that fits this analogy is *soldier* and *warrior* because both terms describe combatants who fight.

Cap is to shoe as jacket is to

    a. Ring.
    b. T-shirt.
    c. Vest.
    d. Glasses.
    e. Pants.

Opposites are at play here because a *cap* is worn on the head/top of the person, while a *shoe* is worn on the foot/bottom. A *jacket* is worn on top of the body too, so the opposite of jacket would be *pants* (*E*) because these are worn on the bottom. Often the prompts on the test provide a synonym or antonym relationship. Just consider if the sets in the prompt reflect similarity or stark difference.

Parts of a Whole

Another thing to consider when first looking at an analogy prompt is whether the words presented come together in some way. Do they express parts of the same item? Does one word complete the other? Are they connected by process or function?

Tire is to car as

> a. Wing is to bird.
> b. Oar is to boat.
> c. Box is to shelf.
> d. Hat is to head.
> e. Knife is to sheath.

We know that the *tire* fits onto the car's wheels and this is what enables the car to drive on roads. The tire is part of the car. This is the same relationship as *oar is to boat* (*B*). The oars are attached onto a boat and enable a person to move and navigate the boat on water. At first glance, *wing is to bird* (*A*) seems to fit too, since a wing is a part of a bird that enables it to move through the air. However, since a tire and car are not alive and transport people, oar and boat fit better because they are also not alive and they transport people. Subtle differences between answer choices should be found.

Other Relationships

There are a number of other relationships to look for when solving verbal analogies. Some relationships focus on one word being a **characteristic/NOT a characteristic** of the other word. Sometimes the first word is the **source/comprised of** the second word. Still, other words are related by their **location**. Some analogies have **sequential** relationships, and some are **cause/effect** relationships. There are analogies that show **creator/provider** relationships with the **creation/provision**. Another relationship might compare an **object with its function** or a **user with his or her tool.** An analogy may focus on a **change of grammar** or a **translation of language**. Finally, one word of an analogy may have a relationship to the other word in its **intensity.**

The type of relationship between the first two words of the analogy should be determined before continuing to analyze the second set of words. One effective method of determining a relationship between two words is to form a comprehensible sentence using both words. Then plug the answer choices into the same sentence. For example in the analogy: *Bicycle is to handlebars as car is to steering wheel*, a sentence could be formed that says: A bicycle navigates using its handlebars; therefore, a car navigates using its steering wheel. If the second sentence makes sense, then the correct relationship likely is found. A sentence may be more complex depending on the relationship between the first two words in the analogy. An example of this may be: *food is to dishwasher as dirt is to carwash.* The formed sentence may be: A dishwasher cleans food off of dishes in the same way that a carwash cleans dirt off of a car.

Dealing with Multiple Connections

There are many other ways to draw connections between word sets. Several word choices might form an analogy that would fit the word set in your prompt. This is when an analogy from multiple angles needs to be explored. Several words might even fit in a relationship. If so, which one is an even closer match than the others? The framing word pair is another important point to consider. Can one or both words be interpreted as actions or ideas, or are they purely objects? Here's a question where words could have alternate meanings:

Hammer is to nail as saw is to

   a. Electric.
   b. Hack.
   c. Cut.
   d. Machete.
   e. Groove.

Looking at the question above, it becomes clear that the topic of the analogy involves construction tools. *Hammers* and *nails* are used in concert, since the hammer is used to pound the nail. The logical first thing to do is to look for an object that a *saw* would be used on. Seeing that there is no such object among the answer choices, a test taker might begin to worry. After all, that seems to be the choice that would complete the analogy—but that doesn't mean it's the only choice that may fit. Encountering questions like this tests the ability to see multiple connections between words. Don't get stuck thinking that words have to be connected in a single way. The first two words given can be verbs instead of just objects. To *hammer* means to hit or beat; oftentimes it refers to beating something into place. This is also what *nail* means when it is used as a verb. Here are the word choices that reveal the answer.

First, it's known that a saw, well, saws. It uses a steady motion to cut an object, and indeed to saw means to cut! *Cut* (C) is one of our answer choices, but the other options should be reviewed. While some tools are *electric* (A), the use of power in the tools listed in the analogy isn't a factor. Again, it's been established that these word choices are not tools in this context. Therefore, *machete* (D) is also ruled out because machete is also not a verb. Another important thing to consider is that while a machete is a tool that accomplishes a similar purpose as a saw, the machete is used in a slicing motion rather than a sawing/cutting motion. The verb that describes machete is *hack* (B), another choice that can be ruled out. A machete is used to hack at foliage. However, a saw does not hack. *Groove* (E) is just a term that has nothing to do with the other words, so this choice can be eliminated easily. This leaves *cut* (C), which confirms that this is the word needed to complete the analogy!

Practice Questions

For the phrases stated below, find their analogy in one of the answer choices.

1. *Cat* is to *paws* as
    a. Giraffe is to neck.
    b. Elephant is to ears.
    c. Horse is to hooves.
    d. Snake is to skin.
    e. Turtle is to shell.

2. *Dancing* is to *rhythm* as *singing* is to
    a. Pitch.
    b. Mouth.
    c. Sound.
    d. Volume.
    e. Words.

3. *Towel* is to *dry* as *hat* is to
    a. Cold.
    b. Warm.
    c. Expose.
    d. Cover.
    e. Top.

4. *Sand* is to *glass* as
    a. Protons are to atoms.
    b. Ice is to snow.
    c. Seeds are to plants.
    d. Water is to steam.
    e. Air is to wind.

5. *Design* is to *create* as *allocate* is to
    a. Finish.
    b. Manage.
    c. Multiply.
    d. Find.
    e. Distribute.

6. *Books* are to *reading* as
    a. Movies are to making.
    b. Shows are to watching.
    c. Poetry is to writing.
    d. Scenes are to performing.
    e. Concerts are to music.

7. *Cool* is to *frigid* as *warm* is to
    a. Toasty.
    b. Summer.
    c. Sweltering.
    d. Hot.
    e. Mild.

8. *Buses* are to *rectangular prisms* as *globes* are to
    a. Circles.
    b. Maps.
    c. Wheels.
    d. Spheres.
    e. Movement.

9. *Backpacks* are to *textbooks* as
    a. Houses are to people.
    b. Fences are to trees.
    c. Plates are to food.
    d. Chalkboards are to chalk.
    e. Computers are to mice.

10. *Storm* is to *rainbow* as *sunset* is to
    a. Clouds.
    b. Sunrise.
    c. Breakfast.
    d. Bedtime.
    e. Stars.

11. *Falcon* is to *mice* as *giraffe* is to
    a. Leaves.
    b. Rocks.
    c. Antelope.
    d. Grasslands.
    e. Hamsters.

12. *Car* is to *motorcycle* as *speedboat* is to
    a. Raft.
    b. Jet-ski.
    c. Sailboat.
    d. Plane.
    e. Canoe.

13. *Arid* is to *damp* as *anxious* is to
    a. Happy.
    b. Petrified.
    c. Ireful.
    d. Confident.
    e. Sorrowful.

14. *Mechanic* is to *repair* as
    a. Mongoose is to cobra.
    b. Rider is to bicycle.
    c. Tree is to grow.
    d. Food is to eaten.
    e. Doctor is to heal.

15. *Whistle* is to *blow horn* as *painting* is to
    a. View.
    b. Criticize.
    c. Sculpture.
    d. Painter.
    e. Paintbrush.

16. *Paddle* is to *boat* as *keys* are to
    a. Unlock.
    b. Success.
    c. Illuminate.
    d. Piano.
    e. Keychain.

17. *Mountain* is to *peak* as *wave* is to
    a. Ocean.
    b. Surf.
    c. Fountain.
    d. Wavelength.
    e. Crest.

18. *Fluent* is to *communication* as
    a. Crater is to catastrophe.
    b. Gourmet is to cooking.
    c. Ink is to pen.
    d. Crow is to raven.
    e. Whistle is to whistler.

19. *Validate* is to *truth* as *conquer* is to
    a. Withdraw.
    b. Subjugate.
    c. Expand.
    d. Surrender.
    e. Expose.

20. *Winter* is to *autumn* as *summer* is to
    a. Vacation.
    b. Fall.
    c. Spring.
    d. March.
    e. Weather.

21. *Fiberglass* is to *surfboard* as
    a. Bamboo is to panda.
    b. Capital is to D.C.
    c. Copper is to penny.
    d. Flint is to mapping.
    e. Wind is to windmill.

22. *Myth* is to *explain* as *joke* is to
    a. Enlighten.
    b. Inspire.
    c. Collect.
    d. Laughter.
    e. Amuse.

23. *Cow* is to *milk* as:
    a. Horse is to cow
    b. Egg is to chicken
    c. Chicken is to egg
    d. Glass is to milk
    e. Milk is to glass

24. *Web* is to *spider* as *den* is to:
    a. Living room
    b. Eagle
    c. Fox
    d. Dog
    e. Turtle

25. *Sad* is to *blue* as *happy* is to:
    a. Glad
    b. Yellow
    c. Smiling
    d. Laugh
    e. Calm

26. *Door* is to *store* as *deal* is to:
    a. Money
    b. Purchase
    c. Sell
    d. Wheel
    e. Market

27. *Dog* is to *veterinarian* as *baby* is to
    a. Daycare
    b. Mother
    c. Puppy
    d. Babysitter
    e. Pediatrician

28. *Clock* is to *time* as:
    a. Ruler is to length
    b. Jet is to speed
    c. Alarm is to sleep
    d. Drum is to beat
    e. Watch is to wrist

29. *Wire* is to *electricity* as:
    a. Power is to lamp
    b. Pipe is to water
    c. Fire is to heat
    d. Heat is to fire
    e. Water is to pipe

30. *President* is to *Executive Branch* as _____ is to *Judicial Branch*.
    a. Supreme Court Justice
    b. Judge
    c. Senator
    d. Lawyer
    e. Congressmen

Answer Explanations

1. C: This is a part/whole analogy. The common thread is what animals walk on. *A, B,* and *E* all describe signature parts of animals, but paws are not the defining feature of cats. While snakes travel on their skins, they do not walk.

2. A: This is a characteristic analogy. The connection lies in what observers will judge a performance on. While the other choices are also important, an off-key singer is as unpleasant as a dancer with no rhythm.

3. D: This is a use/tool analogy. The analogy focuses on an item's use. While hats are worn when it's cold with the goal of making the top of your head warm, this is not always guaranteed—their primary use is to provide cover. There is also the fact that not all hats are used to keep warm, but all hats cover the head.

4. D: This is a source/comprised of analogy. The common thread is addition of fire. Protons contribute to atoms and seeds grow into plants, but these are simple matters of building and growing without necessarily involving fire. Choices *B* and *E* relate objects that already have similar properties.

5. E: This is a synonym analogy. The determining factor is synonymous definition. *Design* and *create* are synonyms, as are *allocate* and *distribute*. Typically, items are found and allocated as part of management to finish a project, but these qualities are not innate in the word. Allocation generally refers to the division of commodities instead of multiplication.

6. B: This is a tool/use analogy. The common thread is audience response to an art form. *A, C,* and *D* deal with the creation of artwork instead of its consumption. Choice *E* describes a form of art instead of the audience engagement with such.

7. C: This is an intensity analogy. The common thread is degree of severity. While *A, D,* and *E* can all describe warmth, they don't convey the harshness of *sweltering*. Choice *B* simply describes a time that people associate with warmth.

8. D: This is a characteristic analogy and is based on matching objects to their geometric shapes. Choice *A* is not correct because globes are three-dimensional, whereas circles exist in two dimensions. While wheels are three-dimensional, they are not always solid or perfectly round.

9. A: This is a tool/use analogy. The key detail of this analogy is the idea of enclosing or sealing items/people. When plates are filled with food, there is no way to enclose the item. While trees can be inside a fence, they can also be specifically outside of one.

10. E: This is a sequence of events analogy. The common thread is celestial cause-and-effect. Not everyone has breakfast or goes to bed after sunset. Sunrise is not typically thought of as the next interesting celestial event after sunsets. While clouds can develop after sunsets, they are also present before and during this activity. Stars, however, can be seen after dark.

11. A: This is a provider/provision analogy. The theme of this analogy is pairing a specific animal to their food source. Falcons prey on mice. Giraffes are herbivores and only eat one of the choices: leaves. Grasslands describe a type of landscape, not a food source for animals.

12. B: This is a category analogy. The common thread is motorized vehicles. While *A, C,* and *E* also describe vehicles that move on water, they are not motorized. Although relying on engines, planes are not a form of water transportation.

13. D: This is an antonym analogy. The prevailing connection is opposite meanings. While *happy* can be an opposite of *anxious,* it's also possible for someone to experience both emotions at once. Choices *B, C,* and *E* are also concurrent with anxious, not opposite.

14. E: This is a provider/provision analogy. This analogy looks at professionals and what their job is. Just as a mechanic's job is to repair machinery, a doctor works to heal patients.

15. C: This is a category analogy. Both whistles and blow horns are devices used to project/produce sound. Therefore, the analogy is based on finding something of a categorical nature. While Choices *A, B, D,* and *E* involve or describe painting, they do not pertain to a distinct discipline alongside painting. Sculpture, however, is another form of art and expression, just like painting.

16. D: This is a part/whole analogy. This analogy examines the relationship between two objects. Specifically, this analogy examines how one object connects to another object, with the first object(s) being the means by which people use the corresponding object to produce a result directly. People use a paddle to steer a boat, just as pressing keys on a piano produces music. Choices *B* and *C* can be metaphorically linked to keys but are unrelated. Choice *A* is related to keys but is a verb, not another object. Choice *E* is the trickiest alternative, but what's important to remember is that while keys are connected to key chains; there is no result just by having the key on a key chain.

17. E: This is a part/whole analogy. This analogy focuses on natural formations and their highest points. The peak of a mountain is its highest point just as the crest is the highest rise in a wave.

18. B: This is an intensity analogy. Fluent refers to how well one can communicate, while gourmet describes a standard of cooking. The analogy draws on degrees of a concept.

19. B: This is a synonym analogy, which relies on matching terms that are most closely connected. *Validate* refers to finding truth. Therefore, finding the term that best fits *conquer* is a good strategy. While nations have conquered others to expand their territory, they are ultimately subjugating those lands and people to their will. Therefore, *subjugate* is the best-fitting answer.

20. C: This is a sequence of events analogy. This analogy pairs one season with the season that precedes it. Winter is paired with autumn because autumn actually comes before winter. Out of all the answers only Choices *B* and *C* are actual seasons. *Fall* is another name for autumn, which comes after summer, not before. Spring, of course, is the season that comes before summer, making it the correct answer.

21. C: This is a source/comprised of analogy. This analogy focuses on pairing a raw material with an object that it's used to create. Fiberglass is used to build surfboards just as copper is used in the creation of pennies. While wind powers a windmill, there is no physical object produced, like with the fiberglass/surfboards pair.

22. E: This is an object/function analogy. The common thread between these words is that one word describes a kind of story and it is paired with the purpose of the story. Myth is/was told in order to explain fundamental beliefs and natural phenomena. While laughter can result from a joke, the purpose of telling a joke is to amuse the audience, thus making Choice *E* the right choice.

23. C: Cows produce milk so the question is looking for another pair that has a producer and their product. Horses don't produce cows (Choice *A*), glasses don't produce milk (Choice *D*), and milk doesn't produce a glass (Choice *E*). The correct choice is *C*: chicken is to egg. The tricky one here is Choice *B*, egg is to chicken, because it has the correct words but the wrong order, and therefore it reverses the relationship. Eggs don't produce chickens, so it doesn't work with the first part of the analogy: cow is to milk.

24. C: The first part of the analogy—web is to spider—describes the home (web) and who lives in it (spider), so the question is looking for what animal lives in a den. The best choice is *C*, fox. Living room is another word or a synonym for a den.

25. A: *Sad* and *blue* are synonyms because they are both describing the same type of mood. The word *blue* in this case is not referring to the color; therefore, although yellow is sometimes considered a "happy" color, the question isn't referring to blue as a color (or any color for that matter) and *yellow* and *happy* are not synonyms. Someone who is happy may laugh or smile, but these words are not synonyms for happy. Lastly, someone who is happy may be calm, although he or she could also be excited, and calm and happy are not synonyms. The best choice is *glad*.

26. D: The key to answering this question correctly is to recognize that the relationship between *door* and *store* is that the words rhyme. One may at first consider the fact that stores have doors, but after reviewing the other word choices and the given word *deal*, he or she should notice that none of the other words have this relationship. Instead, the answer choice should be one that rhymes with *deal*. *Wheel* and *deal*, although spelled differently, are rhyming words, and therefore the correct answer is *D*.

27. E: This question relies on the test takers knowledge of occupations. Dogs are taken care of by veterinarians so the solution is looking for who takes care of babies. However, this level of detail is not yet specific enough because mothers and babysitters can also take care of babies. Veterinarians take care of sick dogs and act as a medical doctor for pets. Therefore, with this higher level of specificity and detail, test takers should select *pediatrician*, because pediatricians are doctors for babies and children.

28. A: The relationship in the first half of the analogy is that clocks are used to measure time, so the second half of the analogy should have a tool that is used to measure something followed by what it measures. Rulers can be used to measure length, so that is the best choice. Remember that the key to solving analogies is to be a good detective. Some of the other answer choices are related to clocks and time but not to the relationship *between* clocks and time.

29. B: Wires are the medium that carry electricity, allowing the current to flow in a circuit. Pipes carry water in a similar fashion, so the best choice is *B*. Test takers must be careful to not select Choice *E*, which reverses the relationship between the components. Choices *A, C,* and *D* contain words that are related to one another but not in the same manner as wires and electricity.

30. A: This question pulls from knowledge in social studies class, understanding the basic roles and positions of the three branches of the government. The president serves in the Executive Branch and the Supreme Court Justices serve in the Judicial Branch.

Reading Comprehension

The Purpose of a Passage

No matter the genre or format, all authors are writing to persuade, inform, entertain, or express feelings. Often, these purposes are blended, with one dominating the rest. It's useful to learn to recognize the author's intent.

Persuasive writing is used to persuade or convince readers of something. It often contains two elements: the argument and the counterargument. The **argument** takes a stance on an issue, while the **counterargument** pokes holes in the opposition's stance. Authors rely on logic, emotion, and writer credibility to persuade readers to agree with them. If readers are opposed to the stance before reading, they are unlikely to adopt that stance. However, those who are undecided or committed to the same stance are more likely to agree with the author.

Informative writing tries to teach or inform. Workplace manuals, instructor lessons, statistical reports, and cookbooks are examples of informative texts. Informative writing is usually based on facts and is often void of emotion and persuasion. Informative texts generally contain statistics, charts, and graphs. Though most informative texts lack a persuasive agenda, readers must examine the text carefully to determine whether one exists within a given passage.

Stories or narratives are designed to entertain. When you go to the movies, you often want to escape for a few hours, not necessarily to think critically. Entertaining writing is designed to delight and engage the reader. However, sometimes this type of writing can be woven into more serious materials, such as persuasive or informative writing to hook the reader before transitioning into a more scholarly discussion.

Emotional writing works to evoke the reader's feelings, such as anger, euphoria, or sadness. The connection between reader and author is an attempt to cause the reader to share the author's intended emotion or tone. Sometimes in order to make a piece more poignant, the author simply wants readers to feel emotion that the author has felt. Other times, the author attempts to persuade or manipulate the reader into adopting his stance. While it's okay to sympathize with the author, be aware of the individual's underlying intent.

Types of Passages

Writing can be classified under four passage types: narrative, expository, descriptive (sometimes called technical), and persuasive. Though these types are not mutually exclusive, one form tends to dominate the rest. By recognizing the *type* of passage you're reading, you gain insight into *how* you should read. If you're reading a narrative, you can assume the author intends to entertain, which means you may skim the text without losing meaning. A technical document might require a close read, because skimming the passage might cause the reader to miss salient details.

1. **Narrative writing**, at its core, is the art of storytelling. For a narrative to exist, certain elements must be present. It must have characters. While many characters are human, **characters** could be defined as anything that thinks, acts, and talks like a human. For example, many recent movies, such as *Lord of the Rings* and *The Chronicles of Narnia*, include animals, fantastical creatures, and even trees that behave like humans. It must have a **plot** or sequence of events. Typically, those events follow a standard plot diagram, but recent trends start *in medias res* or in the middle (near the climax). In this instance,

foreshadowing and flashbacks often fill in plot details. Along with characters and a plot, there must also be conflict. **Conflict** is usually divided into two types: internal and external. **Internal conflict** indicates the character is in turmoil. Internal conflicts are presented through the character's thoughts. **External conflicts** are visible. Types of external conflict include a person versus nature, another person, and society.

2. **Expository writing** is detached and to the point. Since expository writing is designed to instruct or inform, it usually involves directions and steps written in second person ("you" voice) and lacks any persuasive or narrative elements. Sequence words such as *first*, *second*, and *third*, or *in the first place*, *secondly*, and *lastly* are often given to add fluency and cohesion. Common examples of expository writing include instructor's lessons, cookbook recipes, and repair manuals.

3. Due to its empirical nature, **technical writing** is filled with steps, charts, graphs, data, and statistics. The goal of technical writing is to advance understanding in a field through the scientific method. Experts such as teachers, doctors, or mechanics use words unique to the profession in which they operate. These words, which often incorporate acronyms, are called **jargon**. Technical writing is a type of expository writing but is not meant to be understood by the general public. Instead, technical writers assume readers have received a formal education in a particular field of study and need no explanation as to what the jargon means. Imagine a doctor trying to understand a diagnostic reading for a car or a mechanic trying to interpret lab results. Only professionals with proper training will fully comprehend the text.

4. **Persuasive writing** is designed to change opinions and attitudes. The topic, stance, and arguments are found in the thesis, positioned near the end of the introduction. Later supporting paragraphs offer relevant quotations, paraphrases, and summaries from primary or secondary sources, which are then interpreted, analyzed, and evaluated. The goal of persuasive writers is not to stack quotes, but to develop original ideas by using sources as a starting point. Good persuasive writing makes powerful arguments with valid sources and thoughtful analysis. Poor persuasive writing is riddled with bias and logical fallacies. Sometimes, logical and illogical arguments are sandwiched together in the same piece. Therefore, readers should display skepticism when reading persuasive arguments.

Text Structure

Depending on what the author is attempting to accomplish, certain formats or text structures work better than others. For example, a sequence structure might work for narration but not when identifying similarities and differences between dissimilar concepts. Similarly, a comparison-contrast structure is not useful for narration. It's the author's job to put the right information in the correct format.

Readers should be familiar with the five main literary structures:

1. **Sequence** structure (sometimes referred to as the **order** structure) is when the order of events proceed in a predictable order. In many cases, this means the text goes through the plot elements: exposition, rising action, climax, falling action, and resolution. Readers are introduced to characters, setting, and conflict in the exposition. In the rising action, there's an increase in tension and suspense. The climax is the height of tension and the point of no return. Tension decreases during the falling action. In the resolution, any conflicts presented in the exposition are solved, and the story concludes. An informative text that is structured sequentially will often go in order from one step to the next.

2. In the **problem-solution** structure, authors identify a potential problem and suggest a solution. This form of writing is usually divided into two paragraphs and can be found in informational texts. For example, cell phone, cable, and satellite providers use this structure in manuals to help customers troubleshoot or identify problems with services or products.

3. When authors want to discuss similarities and differences between separate concepts, they arrange thoughts in a **comparison-contrast** paragraph structure. Venn diagrams are an effective graphic organizer for comparison-contrast structures, because they feature two overlapping circles that can be used to organize similarities and differences. A comparison-contrast essay organizes one paragraph based on similarities and another based on differences. A comparison-contrast essay can also be arranged with the similarities and differences of individual traits addressed within individual paragraphs. Words such as *however*, *but*, and *nevertheless* help signal a contrast in ideas.

4. **Descriptive** writing structure is designed to appeal to your senses. Much like an artist who constructs a painting, good descriptive writing builds an image in the reader's mind by appealing to the five senses: sight, hearing, taste, touch, and smell. However, overly descriptive writing can become tedious; sparse descriptions can make settings and characters seem flat. Good authors strike a balance by applying descriptions only to passages, characters, and settings that are integral to the plot.

5. Passages that use the **cause and effect** structure are simply asking *why* by demonstrating some type of connection between ideas. Words such as *if*, *since*, *because*, *then*, or *consequently* indicate relationship. By switching the order of a complex sentence, the writer can rearrange the emphasis on different clauses. Saying *If Sheryl is late, we'll miss the dance* is different from saying *We'll miss the dance if Sheryl is late*. One emphasizes Sheryl's tardiness while the other emphasizes missing the dance. Paragraphs can also be arranged in a cause and effect format. Since the format—before and after—is sequential, it is useful when authors wish to discuss the impact of choices. Researchers often apply this paragraph structure to the scientific method.

Point of View

Point of view is an important writing device to consider. In fiction writing, point of view refers to who tells the story or from whose perspective readers are observing as they read. In nonfiction writing, the point of view refers to whether the author refers to himself/herself, his/her readers, or chooses not to refer to either. Whether fiction or nonfiction, the author will carefully consider the impact the perspective will have on the purpose and main point of the writing.

- **First-person point of view:** The story is told from the writer's perspective. In fiction, this would mean that the main character is also the narrator. First-person point of view is easily recognized by the use of personal pronouns such as *I*, *me*, *we*, *us*, *our*, *my*, and *myself*.

- **Third-person point of view:** In a more formal essay, this would be an appropriate perspective because the focus should be on the subject matter, not the writer or the reader. Third-person point of view is recognized by the use of the pronouns *he*, *she*, *they*, and *it*. In fiction writing, third person point of view has a few variations.

o **Third-person limited** point of view refers to a story told by a narrator who has access to the thoughts and feelings of just one character.

o In **third-person omniscient** point of view, the narrator has access to the thoughts and feelings of all the characters.

o   In **third-person objective** point of view, the narrator is like a fly on the wall and can see and hear what the characters do and say but does not have access to their thoughts and feelings.

- **Second-person point of view**: This point of view isn't commonly used in fiction or nonfiction writing because it directly addresses the reader using the pronouns *you*, *your*, and *yourself*. Second-person perspective is more appropriate in direct communication, such as business letters or emails.

Point of View	Pronouns Used
First person	I, me, we, us, our, my, myself
Second person	You, your, yourself
Third person	He, she, it, they

Main Ideas and Supporting Details

Topics and main ideas are critical parts of writing. The **topic** is the subject matter of the piece. An example of a topic would be *the use of cell phones in a classroom*.

The **main idea** is what the writer wants to say about that topic. A writer may make the point that the use of cell phones in a classroom is a serious problem that must be addressed in order for students to learn better. Therefore, the topic is cell phone usage in a classroom, and the main idea is that *it's a serious problem needing to be addressed*. The topic can be expressed in a word or two, but the main idea should be a complete thought.

An author will likely identify the topic immediately within the title or the first sentence of the passage. The main idea is usually presented in the introduction. In a single passage, the main idea may be identified in the first or last sentence, but it will most likely be directly stated and easily recognized by the reader. Because it is not always stated immediately in a passage, it's important that readers carefully read the entire passage to identify the main idea.

The main idea should not be confused with the thesis statement. A **thesis statement** is a clear statement of the writer's specific stance and can often be found in the introduction of a nonfiction piece. The thesis is a specific sentence (or two) that offers the direction and focus of the discussion.

In order to illustrate the main idea, a writer will use **supporting details**, which provide evidence or examples to help make a point. Supporting details are typically found in nonfiction pieces that seek to inform or persuade the reader.

In the example of cell phone usage in the classroom, where the author's main idea is to show the seriousness of this problem and the need to "unplug," supporting details would be critical for effectively making that point. Supporting details used here might include statistics on a decline in student focus and studies showing the impact of digital technology usage on students' attention spans. The author could also include testimonies from teachers surveyed on the topic.

It's important that readers evaluate the author's supporting details to be sure that they are credible, provide evidence of the author's point, and directly support the main idea. Although shocking statistics grab readers' attention, their use may provide ineffective information in the piece. Details like this are crucial to understanding the passage and evaluating how well the author presents his or her argument and evidence.

Also remember that when most authors write, they want to make a point or send a message. This point or message of a text is known as the theme. Authors may state themes explicitly, like in *Aesop's Fables*. More often, especially in modern literature, readers must infer the theme based on text details. Usually after carefully reading and analyzing an entire text, the theme emerges. Typically, the longer the piece, the more themes you will encounter, though often one theme dominates the rest, as evidenced by the author's purposeful revisiting of it throughout the passage.

Evaluating a Passage

Determining conclusions requires being an active reader, as a reader must make a prediction and analyze facts to identify a conclusion. There are a few ways to determine a logical conclusion, but careful reading is the most important. It's helpful to read a passage a few times, noting details that seem important to the piece. A reader should also identify key words in a passage to determine the logical conclusion or determination that flows from the information presented.

Textual evidence within the details helps readers draw a conclusion about a passage. **Textual evidence** refers to information—facts and examples that support the main point. Textual evidence will likely come from outside sources and can be in the form of quoted or paraphrased material. In order to draw a conclusion from evidence, it's important to examine the credibility and validity of that evidence as well as how (and if) it relates to the main idea.

If an author presents a differing opinion or a **counterargument** in order to refute it, the reader should consider how and why this information is being presented. It is meant to strengthen the original argument and shouldn't be confused with the author's intended conclusion, but it should also be considered in the reader's final evaluation.

Sometimes, authors explicitly state the conclusion they want readers to understand. Alternatively, a conclusion may not be directly stated. In that case, readers must rely on the implications to form a logical conclusion:

> On the way to the bus stop, Michael realized his homework wasn't in his backpack. He ran back to the house to get it and made it back to the bus just in time.

In this example, though it's never explicitly stated, it can be inferred that Michael is a student on his way to school in the morning. When forming a conclusion from implied information, it's important to read the text carefully to find several pieces of evidence in the text to support the conclusion.

Summarizing is an effective way to draw a conclusion from a passage. A **summary** is a shortened version of the original text, written by the reader in his/her own words. Focusing on the main points of the original text and including only the relevant details can help readers reach a conclusion. It's important to retain the original meaning of the passage.

Like summarizing, **paraphrasing** can also help a reader fully understand different parts of a text. Paraphrasing calls for the reader to take a small part of the passage and list or describe its main points. Paraphrasing is more than rewording the original passage, though. It should be written in the reader's own words, while still retaining the meaning of the original source. This will indicate an understanding of the original source, yet still help the reader expand on his/her interpretation.

Readers should pay attention to the **sequence**, or the order in which details are laid out in the text, as this can be important to understanding its meaning as a whole. Writers will often use transitional words to help the reader understand the order of events and to stay on track. Words like *next, then, after*, and *finally* show that the order of events is important to the author. In some cases, the author omits these transitional words, and the sequence is implied. Authors may even purposely present the information out of order to make an impact or have an effect on the reader. An example might be when a narrative writer uses **flashback** to reveal information.

Responding to a Passage

There are a few ways for readers to engage actively with the text, such as making inferences and predictions. An **inference** refers to a point that is implied (as opposed to directly-stated) by the evidence presented:

> Bradley packed up all of the items from his desk in a box and said goodbye to his coworkers for the last time.

From this sentence, though it is not directly stated, readers can infer that Bradley is leaving his job. It's necessary to use inference in order to draw conclusions about the meaning of a passage. Authors make implications through character dialogue, thoughts, effects on others, actions, and looks. Like in life, readers must assemble all the clues to form a complete picture.

When making an inference about a passage, it's important to rely only on the information that is provided in the text itself. This helps readers ensure that their conclusions are valid.

Readers will also find themselves making predictions when reading a passage or paragraph. **Predictions** are guesses about what's going to happen next. Readers can use prior knowledge to help make accurate predictions. Prior knowledge is best utilized when readers make links between the current text, previously read texts, and life experiences. Some texts use suspense and foreshadowing to captivate readers:

> A cat darted across the street just as the car came careening around the curve.

One unfortunate prediction might be that the car will hit the cat. Of course, predictions aren't always accurate, so it's important to read carefully to the end of the text to determine the accuracy of predictions.

Critical Thinking Skills

It's important to read any piece of writing critically. The goal is to discover the point and purpose of what the author is writing about through analysis. It's also crucial to establish the point or stance the author has taken on the topic of the piece. After determining the author's perspective, readers can then more effectively develop their own viewpoints on the subject of the piece.

It is important to distinguish between fact and opinion when reading a piece of writing. A **fact** is information that can be proven true. If information can be disproved, it is not a fact. For example, water freezes at or below thirty-two degrees Fahrenheit. An argument stating that water freezes at seventy degrees Fahrenheit cannot be supported by data and is therefore not a fact. Facts tend to be associated with science, mathematics, and statistics. **Opinions** are information open to debate. Opinions are often tied to subjective concepts like equality, morals, and rights. They can also be controversial.

Authors often use words like *think, feel, believe,* or *in my opinion* when expressing opinion, but these words won't always appear in an opinion piece, especially if it is formally written. An author's opinion may be backed up by facts, which gives it more credibility, but that opinion should not be taken as fact. A critical reader should be suspect of an author's opinion, especially if it is only supported by other opinions.

Fact	Opinion
There are 9 innings in a game of baseball.	Baseball games run too long.
James Garfield was assassinated on July 2, 1881.	James Garfield was a good president.
McDonalds has stores in 118 countries.	McDonalds has the best hamburgers.

Critical readers examine the facts used to support an author's argument. They check the facts against other sources to be sure those facts are correct. They also check the validity of the sources used to be sure those sources are credible, academic, and/or peer- reviewed. Consider that when an author uses another person's opinion to support his or her argument, even if it is an expert's opinion, it is still only an opinion and should not be taken as fact. A strong argument uses valid, measurable facts to support ideas. Even then, the reader may disagree with the argument as it may be rooted in his or her personal beliefs.

An authoritative argument may use the facts to sway the reader. In the example of global warming, many experts differ in their opinions of what alternative fuels can be used to aid in offsetting it. Because of this, a writer may choose to only use the information and expert opinion that supports his or her viewpoint.

If the argument is that wind energy is the best solution, the author will use facts that support this idea. That same author may leave out relevant facts on solar energy. The way the author uses facts can influence the reader, so it's important to consider the facts being used, how those facts are being presented, and what information might be left out.

Critical readers should also look for errors in the argument such as logical fallacies and bias. A **logical fallacy** is a flaw in the logic used to make the argument. Logical fallacies include slippery slope, straw man, and begging the question. Authors can also reflect **bias** if they ignore an opposing viewpoint or present their side in an unbalanced way. A strong argument considers the opposition and finds a way to refute it. Critical readers should look for an unfair or one-sided presentation of the argument and be skeptical, as a bias may be present. Even if this bias is unintentional, if it exists in the writing, the reader should be wary of the validity of the argument.

Readers should also look for the use of **stereotypes**, which refer to specific groups. Stereotypes are often negative connotations about a person or place and should always be avoided. When a critical reader finds stereotypes in a piece of writing, they should immediately be critical of the argument and consider the validity of anything the author presents. Stereotypes reveal a flaw in the writer's thinking and may suggest a lack of knowledge or understanding about the subject.

Practice Questions

Read the statement or passage and then choose the best answer to the question. Answer the question based on what is stated or implied in the statement or passage.

1. While scientists aren't entirely certain why tornadoes form, they have some clues into the process. Tornadoes are dangerous funnel clouds that occur during a large thunderstorm. When warm, humid air near the ground meets cold, dry air from above, a column of the warm air can be drawn up into the clouds. Winds at different altitudes blowing at different speeds make the column of air rotate. As the spinning column of air picks up speed, a funnel cloud is formed. This funnel cloud moves rapidly and haphazardly. Rain and hail inside the cloud cause it to touch down, creating a tornado. Tornadoes move in a rapid and unpredictable pattern, making them extremely destructive and dangerous. Scientists continue to study tornadoes to improve radar detection and warning times.

The main purpose of this passage is to:
    a. Show why tornadoes are dangerous
    b. Explain how a tornado forms
    c. Compare thunderstorms to tornadoes
    d. Explain what to do in the event of a tornado

2. There are two major kinds of cameras on the market right now for amateur photographers. Camera enthusiasts can either purchase a digital single-lens reflex camera (DSLR) camera or a compact system camera (CSC). The main difference between a DSLR and a CSC is that the DSLR has a full-sized sensor, which means it fits in a much larger body. The CSC uses a mirrorless system, which makes for a lighter, smaller camera. While both take quality pictures, the DSLR generally has better picture quality due to the larger sensor. CSCs still take very good quality pictures and are more convenient to carry than a DSLR. This makes the CSC an ideal choice for the amateur photographer looking to step up from a point-and-shoot camera.

The main difference between the DSLR and CSC is:
    a. The picture quality is better in the DSLR.
    b. The CSC is less expensive than the DSLR.
    c. The DSLR is a better choice for amateur photographers.
    d. The DSLR's larger sensor makes it a bigger camera than the CSC.

3. When selecting a career path, it's important to explore the various options available. Many students entering college may shy away from a major because they don't know much about it. For example, many students won't opt for a career as an actuary, because they aren't exactly sure what it entails. They would be missing out on a career that is very lucrative and in high demand. Actuaries work in the insurance field and assess risks and premiums. The average salary of an actuary is $100,000 per year. Another career option students may avoid, due to lack of knowledge of the field, is a hospitalist. This is a physician that specializes in the care of patients in a hospital, as opposed to those seen in private practices. The average salary of a hospitalist is upwards of $200,000. It pays to do some digging and find out more about these lesser-known career fields.

An actuary is:
    a. A doctor who works in a hospital
    b. The same as a hospitalist
    c. An insurance agent who works in a hospital
    d. A person who assesses insurance risks and premiums

4. Many people are unsure of exactly how the digestive system works. Digestion begins in the mouth where teeth grind up food and saliva breaks it down, making it easier for the body to absorb. Next, the food moves to the esophagus, and it is pushed into the stomach. The stomach is where food is stored and broken down further by acids and digestive enzymes, preparing it for passage into the intestines. The small intestine is where the nutrients are absorbed from food and passed into the bloodstream. Other essential organs like the liver, gall bladder, and pancreas aid the stomach in breaking down food and absorbing nutrients. Finally, food waste is passed into the large intestine where it is eliminated by the body.

The purpose of this passage is to:
    a. Explain how the liver works.
    b. Show why it is important to eat healthy foods
    c. Explain how the digestive system works
    d. Show how nutrients are absorbed by the small intestine

5. Hard water occurs when rainwater mixes with minerals from rock and soil. Hard water has a high mineral count, including calcium and magnesium. The mineral deposits from hard water can stain hard surfaces in bathrooms and kitchens as well as clog pipes. Hard water can stain dishes, ruin clothes, and reduce the life of any appliances it touches, such as hot water heaters, washing machines, and humidifiers.

One solution is to install a water softener to reduce the mineral content of water, but this can be costly. Running vinegar through pipes and appliances and using vinegar to clean hard surfaces can also help with mineral deposits.

From this passage, it can be concluded that:
    a. Hard water can cause a lot of problems for homeowners.
    b. Calcium is good for pipes and hard surfaces.
    c. Water softeners are easy to install.
    d. Vinegar is the only solution to hard water problems.

6. Osteoporosis is a medical condition that occurs when the body loses bone or makes too little bone tissue. This can lead to brittle, fragile bones that easily break. Bones are already porous, and when osteoporosis sets in, the spaces in bones become much larger, causing them to weaken. Both men and women can develop osteoporosis, though it is most common in women over age 50. Loss of bone can be silent and progressive, so it is important to be proactive in prevention of the disease.

The main purpose of this passage is to:
    a. Discuss some of the ways people contract osteoporosis
    b. Describe different treatment options for those with osteoporosis
    c. Explain how to prevent osteoporosis
    d. Define osteoporosis

7. Vacationers looking for a perfect experience should opt out of Disney parks and try a trip on Disney Cruise Lines. While a park offers rides, characters, and show experiences, it also includes long lines, often very hot weather, and enormous crowds. A Disney Cruise, on the other hand, is a relaxing, luxurious vacation that includes many of the same experiences as the parks, minus the crowds and lines. The cruise has top-notch food, maid service, water slides, multiple pools, Broadway-quality shows, and daily character experiences for kids. There are also many activities, such as bingo, trivia contests, and dance parties that can entertain guests of all ages. The cruise even stops at Disney's private island for a beach barbecue with characters, waterslides, and water sports. Those looking for the Disney experience without the hassle should book a Disney cruise.

The main purpose of this passage is to:
    a. Explain how to book a Disney cruise
    b. Show what Disney parks have to offer
    c. Show why Disney parks are expensive
    d. Compare Disney parks to the Disney cruise

8. Coaches of kids' sports teams are increasingly concerned about the behavior of parents at games. Parents are screaming and cursing at coaches, officials, players, and other parents. Physical fights have even broken out at games. Parents need to be reminded that coaches are volunteers, giving up their time and energy to help kids develop in their chosen sport. The goal of kids' sports teams is to learn and develop skills, but it's also to have fun. When parents are out of control at games and practices, it takes the fun out of the sport.

From this passage, it can be concluded that:
    a. Coaches are modeling good behavior for kids.
    b. Organized sports are not good for kids.
    c. Parents' behavior at their kids' games needs to change.
    d. Parents and coaches need to work together.

9. As summer approaches, drowning incidents will increase. Drowning happens very quickly and silently. Most people assume that drowning is easy to spot, but a person who is drowning doesn't make noise or wave his arms. Instead, he will have his head back and his mouth open, with just his face out of the water. A person who is truly in danger of drowning is not able to wave his arms in the air or move much at all. Recognizing these signs of drowning can prevent tragedy.

The main purpose of this passage is to:
    a. Explain the dangers of swimming
    b. Show how to identify the signs of drowning
    c. Explain how to be a lifeguard
    d. Compare the signs of drowning

10. Technology has been invading cars for the last several years, but there are some new high-tech trends that are pretty amazing. It is now standard in many car models to have a rear-view camera, hands-free phone and text, and a touch screen digital display. Music can be streamed from a paired cell phone, and some displays can even be programmed with a personal photo. Sensors beep to indicate there is something in the driver's path when reversing and changing lanes. Rain-sensing windshield wipers and lights are automatic, leaving the driver with little to do but watch the road and enjoy the ride. The next wave of technology will include cars that automatically parallel park, and a self-driving car is on the horizon. These technological advances make it a good time to be a driver.

It can be concluded from this paragraph that:
    a. Technology will continue to influence how cars are made.
    b. Windshield wipers and lights are always automatic.
    c. It is standard to have a rear-view camera in all cars.
    d. Technology has reached its peak in cars.

11. The Brookside area is an older part of Kansas City, developed mainly in the 1920s and 30s, and is considered one of the nation's first "planned" communities with shops, restaurants, parks, and churches all within a quick walk. A stroll down any street reveals charming two-story Tudor and Colonial homes with smaller bungalows sprinkled throughout the beautiful tree-lined streets. It is common to see lemonade stands on the corners and baseball games in the numerous "pocket" parks tucked neatly behind rows of well-manicured houses. The Brookside shops on 63rd street between Wornall Road and Oak Street are a hub of commerce and entertainment where residents freely shop and dine with their pets (and children) in town. This is also a common "hangout" spot for younger teenagers because it is easily accessible by bike for most. In short, it is an idyllic neighborhood just minutes from downtown Kansas City.

Which of the following states the main idea of this paragraph?
    a. The Brookside shops are a popular hangout for teenagers.
    b. There are a number of pocket parks in the Brookside neighborhood.
    c. Brookside is a great place to live.
    d. Brookside has a high crime rate.

12. At its easternmost part, Long Island opens like the upper and under jaws of some prodigious alligator; the upper and larger one terminating in Montauk Point. The bay that lies in here, and part of which forms the splendid harbor of Greenport, where the Long Island Railroad ends, is called Peconic Bay; and a beautiful and varied water is it, fertile in fish and feathered game. I, who am by no means a skillful fisherman, go down for an hour of a morning on one of the docks, or almost anywhere along shore, and catch a mess of black-fish, which you couldn't buy in New York for a dollar—large fat fellows, with meat on their bones that it takes a pretty long fork to stick through. They have a way here of splitting these fat black-fish and poggies, and broiling them on the coals, beef-steak-fashion, which I recommend your Broadway cooks to copy.

Which of the following is the best summary of this passage?
    a. Walt Whitman was impressed with the quantity and quality of fish he found in Peconic Bay.
    b. Walt Whitman prefers the fish found in restaurants in New York.
    c. Walt Whitman was a Broadway chef.
    d. Walt Whitman is frustrated because he is not a very skilled fisherman.

13. Evidently, our country has overlooked both the importance of learning history and the appreciation of it. But why is this a huge problem? Other than historians, who really cares how the War of 1812 began, or who Alexander the Great's tutor was? Well, not many, as it turns out. So, is history really that important? Yes! History is critical to help us understand the underlying forces that shape decisive events, to prevent us from making the same mistakes twice, and to give us context for current events.

The above is an example of which type of writing?
    a. Expository
    b. Persuasive
    c. Narrative
    d. Poetry

14. Although many Missourians know that Harry S. Truman and Walt Disney hailed from their great state, probably far fewer know that it was also home to the remarkable George Washington Carver. At the end of the Civil War, Moses Carver, the slave owner who owned George's parents, decided to keep George and his brother and raise them on his farm. As a child, George was driven to learn and he loved painting. He even went on to study art while in college but was encouraged to pursue botany instead. He spent much of his life helping others by showing them better ways to farm; his ideas improved agricultural productivity in many countries. One of his most notable contributions to the newly emerging class of Negro farmers was to teach them the negative effects of agricultural monoculture, i.e. growing the same crops in the same fields year after year, depleting the soil of much needed nutrients and resulting in a lesser yielding crop. Carver was an innovator, always thinking of new and better ways to do things, and is most famous for his over three hundred uses for the peanut. Toward the end of his career, Carver returned to his first love of art. Through his artwork, he hoped to inspire people to see the beauty around them and to do great things themselves. When Carver died, he left his money to help fund ongoing agricultural research. Today, people still visit and study at the George Washington Carver Foundation at Tuskegee Institute.

According to the passage, what was George Washington Carver's first love?
    a. Plants
    b. Art
    c. Animals
    d. Soil

15. *Please read the following two passages and then answer the question that follows.*

Passage 1

In the modern classroom, cell phones have become indispensable. Cell phones, which are essentially handheld computers, allow students to take notes, connect to the web, perform complex computations, teleconference, and participate in surveys. Most importantly, though, due to their mobility and excellent reception, cell phones are necessary in emergencies. Unlike tablets, laptops, or computers, cell phones are a readily available and free resource—most school district budgets are already strained to begin with—and since today's student is already strongly rooted in technology, when teachers incorporate cell phones, they're "speaking" the student's language, which increases the chance of higher engagement.

Passage 2

As with most forms of technology, there is an appropriate time and place for the use of cell phones. Students are comfortable with cell phones, so it makes sense when teachers allow cell phone use at their discretion. Allowing cell phone use can prove advantageous if done correctly. Unfortunately, if that's not the case—and often it isn't—then a sizable percentage of students pretend to pay attention while surreptitiously playing on their phones. This type of disrespectful behavior is often justified by the argument that cell phones are not only a privilege but also a right. Under this logic, confiscating phones is akin to rummaging through students' backpacks. This is in stark contrast to several decades ago when teachers regulated where and when students accessed information.

With which of the following statements would both the authors of Passages 1 and 2 agree?
    a. Teachers should incorporate cell phones into curriculum whenever possible.
    b. Cell phones are useful only when an experienced teacher uses them properly.
    c. Cell phones and, moreover, technology, are a strong part of today's culture.
    d. Despite a good lesson plan, cell phone disruptions are impossible to avoid.

16. Annabelle Rice started having trouble sleeping. Her biological clock was suddenly amiss and she began to lead a nocturnal schedule. She thought her insomnia was due to spending nights writing a horror story, but then she realized that even the idea of going outside into the bright world scared her to bits. She concluded she was now suffering from heliophobia.

Which of the following most accurately describes the meaning of the underlined word in the sentence above?
    a. Fear of dreams
    b. Fear of sunlight
    c. Fear of strangers
    d. Generalized anxiety disorder

17. A famous children's author recently published a historical fiction novel under a pseudonym; however, it did not sell as many copies as her children's books. In her earlier years, she had majored in history and earned a graduate degree in Antebellum American History, which is the timeframe of her new novel. Critics praised this newest work far more than the children's series that made her famous. In fact, her new novel was nominated for the prestigious Albert J. Beveridge Award, but still isn't selling like her children's books, which fly off the shelves because of her name alone.

Which one of the following statements might be accurately inferred based on the above passage?
    a. The famous children's author produced an inferior book under her pseudonym.
    b. The famous children's author is the foremost expert on Antebellum America.
    c. The famous children's author did not receive the bump in publicity for her historical novel that it would have received if it were written under her given name.
    d. People generally prefer to read children's series than historical fiction.

18. In 2015, 28 countries, including Estonia, Portugal, Slovenia, and Latvia, scored significantly higher than the United States on standardized high school math tests. In the 1960s, the United States consistently ranked first in the world. Today, the United States spends more than $800 billion dollars on education, which exceeds the next highest country by more than $600 billion dollars. The United States also leads the world in spending per school-aged child by an enormous margin.

If these statements above are factual, which of the following statements must be correct?
    a. Outspending other countries on education has benefits beyond standardized math tests.
    b. The United States' education system is corrupt and broken.
    c. The standardized math tests are not representative of American academic prowess.
    d. Spending more money does not guarantee success on standardized math tests.

19. The following exchange occurred after the Baseball Coach's team suffered a heartbreaking loss in the final inning.

Reporter: The team clearly did not rise to the challenge. I'm sure that getting zero hits in twenty at-bats with runners in scoring position hurt the team's chances at winning the game. What are your thoughts on this devastating loss?

Baseball Coach: Hitting with runners in scoring position was not the reason we lost this game. We made numerous errors in the field, and our pitchers gave out too many free passes. Also, we did not even need a hit with runners in scoring position. Many of those at-bats could have driven in the run by simply making contact. Our team did not deserve to win the game.

Which of the following best describes the main point of dispute between the reporter and baseball coach?
    a. Whether the loss was heartbreaking
    b. Whether getting zero hits in twenty at-bats with runners in scoring position caused the loss
    c. That numerous errors in the field and pitchers giving too many free passes caused the loss
    d. Whether the team deserved to win the game

20. Last week, we adopted a dog from the local animal shelter, after looking for our perfect pet for several months. We wanted a dog that was not too old, but also past the puppy stage, so that training would be less time-intensive and so that we would give an older animal a home. Robin, as she's called, was a perfect match and we filled out our application and upon approval, were permitted to bring her home. Her physical exam and lab work all confirmed she was healthy. We went to the pet store and bought all sorts of bedding, food, toys, and treats to outfit our house as a dog-friendly and fun place. The shelter told us she liked dry food only, which is a relief because wet food is expensive and pretty off-putting. We even got fencing and installed a dog run in the backyard for Robin to roam unattended. Then we took her to the vet to make sure she was healthy. Next week, she starts the dog obedience class that we enrolled her in with a discount coupon from the shelter. It will be a good opportunity to bond with her and establish commands and dominance. When we took her to the park the afternoon after we adopted her, it was clear that she is a sociable and friendly dog, easily playing cohesively with dogs of all sizes and dispositions.

Which of the following is out of sequence in the story?
    a. Last week, we adopted a dog from the local animal shelter, after looking for our perfect pet for several months.
    b. Robin, as she's called, was a perfect match and we filled out our application and upon approval, were permitted to bring her home.
    c. Her physical exam and lab work all confirmed she was healthy.
    d. Next week, she starts the dog obedience class that we enrolled her in with a discount coupon from the shelter.

Answer Explanations

1. B: The main point of this passage is to show how a tornado forms. Choice *A* is incorrect because while the passage does mention that tornadoes are dangerous, it is not the main focus of the passage. While thunderstorms are mentioned, they are not compared to tornadoes, so Choice *C* is incorrect. Choice *D* is incorrect because the passage does not discuss what to do in the event of a tornado.

2. D: The passage directly states that the larger sensor is the main difference between the two cameras. Choices *A* and *B* may be true, but these answers do not identify the major difference between the two cameras. Choice *C* states the opposite of what the paragraph suggests is the best option for amateur photographers, so it is incorrect.

3. D: An actuary assesses risks and sets insurance premiums. While an actuary does work in insurance, the passage does not suggest that actuaries have any affiliation with hospitalists or working in a hospital, so all other choices are incorrect.

4. C: The purpose of this passage is to explain how the digestive system works. Choice *A* focuses only on the liver, which is a small part of the process and not the focus of the paragraph. Choice *B* is off-track because the passage does not mention healthy foods. Choice *D* only focuses on one part of the digestive system.

5. A: The passage focuses mainly on the problems of hard water. Choice *B* is incorrect because calcium is not good for pipes and hard surfaces. The passage does not say anything about whether water softeners are easy to install, so *C* is incorrect. Choice *D* is also incorrect because the passage does offer other solutions besides vinegar.

6. D: The main point of this passage is to define osteoporosis. Choice *A* is incorrect because the passage does not list ways that people contract osteoporosis. Choice *B* is incorrect because the passage does not mention any treatment options. While the passage does briefly mention prevention, it does not explain how, so Choice *C* is incorrect.

7. D: The passage compares Disney cruises with Disney parks. It does not discuss how to book a cruise, so Choice *A* is incorrect. Choice *B* is incorrect because though the passage does mention some of the park attractions, it is not the main point. The passage does not mention the cost of either option, so Choice *C* is also incorrect.

8. C: The main point of this paragraph is that parents need to change their poor behavior at their kids' sporting events. Choice *A* is incorrect because the coaches' behavior is not mentioned in the paragraph. Choice *B* suggests that sports are bad for kids, when the paragraph is about parents' behavior, so it is incorrect. While Choice *D* may be true, it offers a specific solution to the problem, which the paragraph does not discuss.

9. B: The point of this passage is to show what drowning looks like. Choice *A* is incorrect because while drowning is a danger of swimming, the passage doesn't include any other dangers. The passage is not intended for lifeguards specifically, but for a general audience, so Choice *C* is incorrect. There are a few signs of drowning, but the passage does not compare them; thus, Choice *D* is incorrect.

10. A: The passage discusses recent technological advances in cars and suggests that this trend will continue in the future with self-driving cars. Choice *B* and *C* are not true, so these are both incorrect. Choice *D* is also incorrect because the passage suggests continuing growth in technology, not a peak.

11. C: All the details in this paragraph suggest that Brookside is a great place to live, plus the last sentence states that it is an *idyllic neighborhood*, meaning it is perfect, happy, and blissful. Choices *A* and *B* are incorrect, because although they do contain specific details from the paragraph that support the main idea, they are not the main idea. Choice *D* is incorrect because there is no reference in the paragraph of the crime rate in Brookside.

12. A: Choice *A* is correct because there is evidence in the passage to support it, specifically when he mentions catching "a mess of black-fish, which you couldn't buy in New York for a dollar—large fat fellows, with meat on their bones that it takes a pretty long fork to stick through." There is no evidence to support the other answer choices.

13. B: Persuasive is the correct answer because the author is clearly trying to convey the point that history education is very important. Choice *A* is incorrect because expository writing is more informative and less emotional. Choice *C* is incorrect because narrative writing involves story telling. Choice *D* is incorrect because this is a piece of prose, not poetry.

14. B: This is the correct answer choice because the passage begins by describing Carver's childhood fascination with painting and later returns to this point when it states that at the end of his career "Carver returned to his first love of art." For this reason, all the other answer choices are incorrect.

15. C: Despite the opposite stances in Passages 1 and 2, both authors establish that cell phones are a strong part of culture. In Passage 1 the author states, "Today's student is already strongly rooted in technology." In Passage 2 the author states, "Students are comfortable with cell phones." The author of Passage 2 states that cell phones have a "time and place." The author of Passage 2 would disagree with the statement that *teachers should incorporate cell phones into curriculum whenever possible. Cell phones are useful only when an experienced teacher uses them properly*—this statement is implied in Passage 2, but the author in Passage 1 says cell phones are "indispensable." In other words, no teacher can do without them. *Despite a good lesson plan, cell phone disruptions are impossible to avoid*. This is not supported by either passage. Even though the author in the second passage is more cautionary, the author states, "This can prove advantageous if done correctly." Therefore, there is a possibility that a classroom can run properly with cell phones.

16. B: The passage indicates that Annabelle has a fear of going outside into the daylight. Thus *heliophobia* must refer to a fear of bright lights or sunlight. Choice *B* is the only answer that describes this.

17. C: We are looking for an inference—a conclusion that is reached on the basis of evidence and reasoning—from the passage that will likely explain why the famous children's author did not achieve her usual success with the new genre (despite the book's acclaim). Choice *A* is wrong because the statement is false according to the passage. Choice *B* is wrong because, although the passage says the author has a graduate degree on the subject, it would be an unrealistic leap to infer that she is the foremost expert on Antebellum America. Choice *D* is wrong because there is nothing in the passage to lead us to infer that people generally prefer a children's series to historical fiction. In contrast, Choice *C* can be logically inferred since the passage speaks of the great success of the children's series and the declaration that the fame of the author's name causes the children's books to "fly off the shelves." Thus, she did not receive any bump from her name since she published the historical novel under a pseudonym, and Choice C is correct.

18. D: Outspending other countries on education could have other benefits, but there is no reference to this in the passage, so Choice A is incorrect. Choice B is incorrect because the author does not mention corruption. Choice C is incorrect because there is nothing in the passage stating that the tests are not genuinely representative. Choice D is accurate because spending more money has not brought success. The United States already spends the most money, and the country is not excelling on these tests. Choice D is the correct answer.

19. B: Choice A uses similar language, but it is not the main point of disagreement. The reporter calls the loss devastating, and there's no reason to believe that the coach would disagree with this assessment. Eliminate this choice.

Choice B is strong since both passages mention the at-bats with runners in scoring position. The reporter asserts that the team lost due to the team failing to get such a hit. In contrast, the coach identifies several other reasons for the loss, including fielding and pitching errors. Additionally, the coach disagrees that the team even needed a hit in those situations.

Choice C is mentioned by the coach, but not by the reporter. It is unclear whether the reporter would agree with this assessment. Eliminate this choice.

Choice D is mentioned by the coach but not by the reporter. It is not stated whether the reporter believes that the team deserved to win. Eliminate this choice.

Therefore, Choice B is the correct answer.

20. C: The passage is told in chronological order; it details the steps the family took to adopt their dog. The narrator mentions that Robin's physical exam and lab work confirmed she was healthy before discussing that they brought her to the vet to evaluate her health. It is illogical that lab work would confirm good health prior to an appointment with the vet, when presumably, the lab work would be collected.

Mathematics

Number Concepts and Operations

Rational Numbers

The mathematical number system is made up of two general types of numbers: real and complex. **Real numbers** are those that are used in normal settings, while **complex numbers** are those composed of both a real number and an imaginary one. Imaginary numbers are the result of taking the square root of -1, and $\sqrt{-1} = i$.

The real number system is often explained using a Venn diagram similar to the one below. After a number has been labeled as a real number, further classification occurs when considering the other groups in this diagram. If a number is a never-ending, non-repeating decimal, it falls in the **irrational** category. Otherwise, it is **rational**. Furthermore, if a number does not have a fractional part, it is classified as an **integer**, such as -2, 75, or 0. **Whole numbers** are an even smaller group that only includes positive integers and 0. The last group of **natural numbers** is made up of only positive integers, such as 2, 56, or 12.

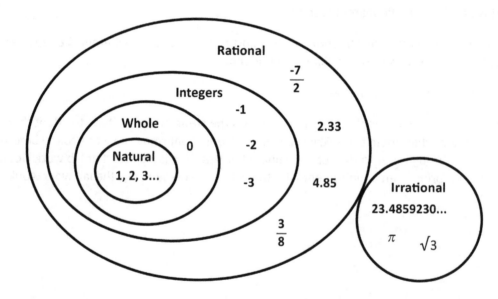

Real numbers can be compared and ordered using the number line. If a number falls to the left on the real number line, it is less than a number on the right. For example, $-2 < 5$ because -2 falls to the left of 0, and 5 falls to the right.

Complex numbers are made up of the sum of a real number and an imaginary number. Some examples of complex numbers include $6 + 2i$, $5 - 7i$, and $-3 + 12i$. Adding and subtracting complex numbers is similar to collecting like terms. The real numbers are added together, and the imaginary numbers are added together. For example, if the problem asks to simplify the given expression $6 + 2i - 3 + 7i$, the 6 and -3 would combine to make 3, and the $2i$ and $7i$ combine to make $9i$. Multiplying and dividing complex numbers is similar to working with exponents. One rule to remember when multiplying is that $i \times i = -1$. For example, if a problem asks to simplify the expression $4i(3 + 7i)$, the $4i$ should be distributed throughout the 3 and the $7i$. This leaves the final expression $12i - 28$. The 28 is negative because $i \times i$ results in a negative number. The last type of operation to consider with complex numbers is the conjugate. The **conjugate** of a complex number is a technique used to change the complex number into a real number. For example, the conjugate of $4 - 3i$ is $4 + 3i$. Multiplying $(4 - 3i)(4 + 3i)$ results in $16 + 12i - 12i + 9$, which has a final answer of $16 + 9 = 25$.

Integers are the whole numbers together with their negatives. They include numbers like 5, 24, 0, -6, and 15. They do not include fractions or numbers that have digits after the decimal point.

Rational numbers are all numbers that can be written as a fraction using integers. A **fraction** is written as $\frac{x}{y}$ and represents the quotient of x being divided by y. More practically, it means dividing the whole into y equal parts, then taking x of those parts.

Examples of rational numbers include $\frac{1}{2}$ and $\frac{5}{4}$. The number on the top is called the **numerator,** and the number on the bottom is called the **denominator**. Because every integer can be written as a fraction with a denominator of 1, (e.g. $\frac{3}{1} = 3$), every integer is also a rational number.

When adding integers and negative rational numbers, there are some basic rules to determine if the solution is negative or positive:

- Adding two positive numbers results in a positive value: $3.3 + 4.8 = 8.1$.
- Adding two negative numbers results in a negative number: $(-8) + (-6) = -14$.

Adding one positive and one negative number requires taking the absolute values and finding the difference between them. Then, the sign of the number that has the higher absolute value for the final solution is used.

For example, $(-9) + 11$, has a difference of absolute values of 2. The final solution is 2 because 11 has the higher absolute value. Another example is $9 + (-11)$, which has a difference of absolute values of 2. The final solution is -2 because 11 has the higher absolute value.

When subtracting integers and negative rational numbers, one has to change the problem to adding the opposite and then apply the rules of addition.

- Subtracting two positive numbers is the same as adding one positive and one negative number. For example, $4.9 - 7.1$ is the same as $4.9 + (-7.1)$. The solution is -2.2 since the absolute value of -7.1 is greater. Another example is $8.5 - 6.4$, which is the same as $8.5 + (-6.4)$. The solution is 2.1 since the absolute value of 8.5 is greater.

- Subtracting a positive number from a negative number results in negative value. For example, $(-12) - 7$ is the same as $(-12) + (-7)$ with a solution of -19.

- Subtracting a negative number from a positive number results in a positive value. For example, $12 - (-7)$ is the same as $12 + 7$ with a solution of 19.

- For multiplication and division of integers and rational numbers, if both numbers are positive or both numbers are negative, the result is a positive value. For example, $(-1.7)(-4)$ has a solution of 6.8 since both numbers are negative values.

- If one number is positive and another number is negative, the result is a negative value. For example, $(-15) \div 5$ has a solution of -3 since there is one negative number.

As mentioned, rational numbers are any number which can be written as a fraction or ratio. Within the set of rational numbers, several subsets exist which are referenced throughout the mathematics topics. **Counting numbers** are the first numbers learned as a child. Counting numbers consist of 1,2,3,4, and so on. **Whole numbers** include all counting numbers and zero (0,1,2,3,4,...). **Integers** include counting numbers, their opposites, and zero (...,-3,-2,-1,0,1,2,3,...). **Rational numbers** are inclusive of integers, fractions, and decimals that terminate, or end (1.7, 0.04213) or repeat ($0.136\overline{5}$).

A **number line** typically consists of integers (...3,2,1,0,-1,-2,-3...), and is used to visually represent the value of a rational number. Each rational number has a distinct position on the line determined by comparing its value with the displayed values on the line. For example, if plotting -1.5 on the number line below, it is necessary to recognize that the value of -1.5 is .5 less than -1 and .5 greater than -2. Therefore, -1.5 is plotted halfway between -1 and -2.

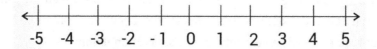

The number system that is used consists of only ten different digits or characters. However, this system is used to represent an infinite number of values. The **place value system** makes this infinite number of values possible. The position in which a digit is written corresponds to a given value. Starting from the decimal point (which is implied, if not physically present), each subsequent place value to the left represents a value greater than the one before it. Conversely, starting from the decimal point, each subsequent place value to the right represents a value less than the one before it.

The names for the place values to the left of the decimal point are as follows:

...	Billions	Hundred-Millions	Ten-Millions	Millions	Hundred-Thousands	Ten-Thousands	Thousands	Hundreds	Tens	Ones

*Note that this table can be extended infinitely further to the left.

The names for the place values to the right of the decimal point are as follows:

Decimal Point (.)	Tenths	Hundredths	Thousandths	Ten-Thousandths	...

*Note that this table can be extended infinitely further to the right.

When given a multi-digit number, the value of each digit depends on its place value. Consider the number 682,174.953. Referring to the chart above, it can be determined that the digit 8 is in the ten-thousands place. It is in the fifth place to the left of the decimal point. Its value is 8 ten-thousands or 80,000. The digit 5 is two places to the right of the decimal point. Therefore, the digit 5 is in the hundredths place. Its value is 5 hundredths or $\frac{5}{100}$ (equivalent to .05).

In accordance with the **base-10 system**, the value of a digit increases by a factor of ten each place it moves to the left. For example, consider the number 7. Moving the digit one place to the left (70), increases its value by a factor of 10 ($7 \times 10 = 70$). Moving the digit two places to the left (700) increases its value by a factor of 10 twice ($7 \times 10 \times 10 = 700$). Moving the digit three places to the left (7,000) increases its value by a factor of 10 three times ($7 \times 10 \times 10 \times 10 = 7,000$), and so on.

Conversely, the value of a digit decreases by a factor of ten each place it moves to the right. (Note that multiplying by $\frac{1}{10}$ is equivalent to dividing by 10). For example, consider the number 40. Moving the digit one place to the right (4) decreases its value by a factor of 10 ($40 \div 10 = 4$). Moving the digit two places to the right (0.4), decreases its value by a factor of 10 twice ($40 \div 10 \div 10 = 0.4$) or ($40 \times \frac{1}{10} \times \frac{1}{10} = 0.4$). Moving the digit three places to the right (0.04) decreases its value by a factor of 10 three times ($40 \div 10 \div 10 \div 10 = 0.04$) or ($40 \times \frac{1}{10} \times \frac{1}{10} \times \frac{1}{10} = 0.04$), and so on.

Ordering Numbers

A common question type asks to order rational numbers from least to greatest or greatest to least. The numbers will come in a variety of formats, including decimals, percentages, roots, fractions, and whole numbers. These questions test for knowledge of different types of numbers and the ability to determine their respective values.

Before discussing ordering all numbers, let's start with decimals.

To compare decimals and order them by their value, utilize a method similar to that of ordering large numbers.

The main difference is where the comparison will start. Assuming that any numbers to left of the decimal point are equal, the next numbers to be compared are those immediately to the right of the decimal point. If those are equal, then move on to compare the values in the next decimal place to the right.

For example:

Which number is greater, 12.35 or 12.38?

Check that the values to the left of the decimal point are equal:

$$12 = 12$$

Next, compare the values of the decimal place to the right of the decimal:

$$12.3 = 12.3$$

Those are also equal in value.

Finally, compare the value of the numbers in the next decimal place to the right on both numbers:

$$12.3\mathbf{5} \text{ and } 12.3\mathbf{8}$$

Here the 5 is less than the 8, so the final way to express this inequality is:

$$12.35 < 12.38$$

Comparing decimals is regularly exemplified with money because the "cents" portion of money ends in the hundredths place. When paying for gasoline or meals in restaurants, and even in bank accounts, if enough errors are made when calculating numbers to the hundredths place, they can add up to dollars and larger amounts of money over time.

Now that decimal ordering has been explained, let's expand and consider all real numbers. Whether the question asks to order the numbers from greatest to least or least to greatest, the crux of the question is the same—convert the numbers into a common format. Generally, it's easiest to write the numbers as whole numbers and decimals so they can be placed on a number line. Follow these examples to understand this strategy.

1) Order the following rational numbers from greatest to least:

$$\sqrt{36}, 0.65, 78\%, \frac{3}{4}, 7, 90\%, \frac{5}{2}$$

Of the seven numbers, the whole number (7) and decimal (0.65) are already in an accessible form, so concentrate on the other five.

First, the square root of 36 equals 6. (If the test asks for the root of a non-perfect root, determine which two whole numbers the root lies between.) Next, convert the percentages to decimals. A percentage means "per hundred," so this conversion requires moving the decimal point two places to the left, leaving 0.78 and 0.9. Lastly, evaluate the fractions:

$$\frac{3}{4} = \frac{75}{100} = 0.75 \,; \frac{5}{2} = 2\frac{1}{2} = 2.5$$

Now, the only step left is to list the numbers in the request order:

$$7, \sqrt{36}, \frac{5}{2}, 90\%, 78\%, \frac{3}{4}, 0.65$$

2) Order the following rational numbers from least to greatest:

$$2.5, \sqrt{9}, -10.5, 0.853, 175\%, \sqrt{4}, \frac{4}{5}$$

$$\sqrt{9} = 3$$

$$175\% = 1.75$$

$$\sqrt{4} = 2$$

$$\frac{4}{5} = 0.8$$

From least to greatest, the answer is:

$$-10.5, \frac{4}{5}, 0.853, 175\%, \sqrt{4}, 2.5, \sqrt{9}$$

Basic Concepts of Addition, Subtraction, Multiplication, and Division

The four most basic and fundamental operations are addition, subtraction, multiplication, and division. The operations take a group of numbers and achieve an exact result, according to the operation. When dealing with positive integers greater than 1, addition and multiplication increase the result, while subtraction and division reduce the result.

Addition

Addition is the combination of two numbers so their quantities are added together cumulatively. The sign for an addition operation is the + symbol. For example, 9 + 6 = 15. The 9 and 6 combine to achieve a cumulative value, called a **sum**.

Addition holds the **commutative property**, which means that numbers in an addition equation can be switched without altering the result. The formula for the commutative property is $a + b = b + a$. Let's look at a few examples to see how the commutative property works:

$$3 + 4 = 4 + 3 = 7$$

$$12 + 8 = 8 + 12 = 20$$

Addition also holds the **associative property**, which means that the groups of numbers don't matter in an addition problem. The formula for the associative property is $(a + b) + c = a + (b + c)$. Here are some examples of the Associative Property at work:

$$(6 + 14) + 10 = 6 + (14 + 10) = 30$$

$$8 + (2 + 25) = (8 + 2) + 25 = 35$$

Subtraction

Subtraction is taking away one number from another, so their quantities are reduced. The sign designating a subtraction operation is the $-$ symbol, and the result is called the **difference**. For example, $9 - 6 = 3$. The second number detracts from the first number in the equation to reach the difference.

Unlike addition, subtraction doesn't follow the commutative or associative properties. The order and grouping in subtraction impacts the result.

$$22 - 7 \neq 7 - 22$$

$$(10 - 5) - 2 \neq 10 - (5 - 2)$$

When working through subtraction problems involving larger numbers, it's necessary to regroup the numbers. Let's work through a practice problem using regrouping:

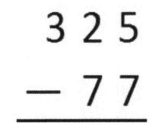

Here, it is clear that the ones and tens columns for 77 are greater than the ones and tens columns for 325. To subtract this number, borrow from the tens and hundreds columns. When borrowing from a column, subtracting 1 from the lender column will add 10 to the borrower column:

$$\begin{array}{ccc} 3\text{-}1 & 10\text{+}2\text{-}1 & 10\text{+}5 \\ - & 7 & 7 \\ \hline \end{array} = \begin{array}{ccc} 2 & 11 & 15 \\ - & 7 & 7 \\ \hline 2 & 4 & 8 \end{array}$$

After ensuring that each digit in the top row is greater than the digit in the corresponding bottom row, subtraction can proceed as normal, and the answer is found to be 248.

Multiplication

Multiplication involves taking multiple copies of one number. The sign designating a multiplication operation is the $\times$ symbol. The result is called the **product**. For example, $9 \times 6 = 54$. Multiplication means adding together one number in the equation as many times as the number on the other side of the equation:

$$9 \times 6 = 9 + 9 + 9 + 9 + 9 + 9 = 54$$

$$9 \times 6 = 6 + 6 + 6 + 6 + 6 + 6 + 6 + 6 + 6 = 54$$

Like addition, multiplication holds the commutative and associative properties:

$$2 \times 4 = 4 \times 2$$

$$2 \times 5 = 5 \times 20$$

$$(2 \times 4) \times 10 = 2 \times (4 \times 10)$$

$$3 \times (7 \times 4) = (3 \times 7) \times 4$$

Multiplication also follows the **distributive property**, which allows the multiplication to be distributed through parentheses. The formula for distribution is $a \times (b + c) = ab + ac$. This will become clearer after some examples:

$$5 \times (3 + 6) = (5 \times 3) + (5 \times 6) = 45$$

$$4 \times (10 - 5) = (4 \times 10) - (4 \times 5) = 20$$

Multiplication becomes slightly more complicated when multiplying numbers with decimals. The easiest way to answer these problems is to ignore the decimals and multiply as if they were whole numbers.

After multiplying the factors, add a decimal point to the product, and the decimal place should be equal to the total number of decimal places in the problem. For example:

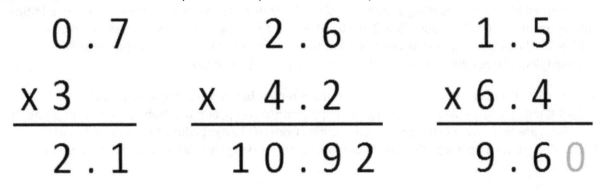

$$
\begin{array}{r} 0.7 \\ \times\,3 \\ \hline 2.1 \end{array}
\qquad
\begin{array}{r} 2.6 \\ \times\,4.2 \\ \hline 10.92 \end{array}
\qquad
\begin{array}{r} 1.5 \\ \times\,6.4 \\ \hline 9.60 \end{array}
$$

Let's tackle the first example. First, ignore the decimal and multiply the numbers as though they were whole numbers to arrive at a product: 21. Second, count the number of digits that follow a decimal (one). Finally, move the decimal place that many positions to the left, as the factors have only one decimal place. The second example works the same way, except that there are two total decimal places in the factors, so the product's decimal is moved two places over. In the third example, the decimal should be moved over two digits, but the digit zero is no longer needed, so it is erased and the final answer is 9.6.

<u>Division</u>
Division is the opposite of multiplication as addition and subtraction are opposites. The signs designating a division operate are the ÷ or / symbols. In division, the second number divides the first. Like subtraction, it matters which number comes before the division sign. For example, $9 \div 6 \neq 6 \div 9$.

The number before the division sign is called the **dividend**, or, if expressed as a fraction, then it's the **numerator**. For example, in a ÷ b, a is the dividend. In $\frac{a}{b}$, a is the numerator.

The number after the division sign is called the **divisor**, or, if expressed as a fraction, then it's the **denominator**. For example, in a ÷ b, b is the divisor. In $\frac{a}{b}$, b is the denominator.

Division doesn't follow any of the commutative, associative, or distributive properties.

$$15 \div 3 \neq 3 \div 15$$

$$(30 \div 3) \div 5 \neq 30 \div (3 \div 5)$$

$$100 \div (20 + 10) \neq (100 \div 20) + (100 \div 10)$$

Division with decimals is similar to multiplication with decimals. When dividing a decimal by a whole number, ignore the decimal and divide as if it were a whole number. Once you solve for the answer (also called the **quotient**), then place the decimal at the decimal place equal to that in the dividend.

$$15.75 \div 3 = 5.25$$

When the divisor is a decimal, change it to a whole number by multiplying both the divisor and dividend by the factor of 10 that makes the divisor into a whole number. Then complete the division operation as described above.

$$17.5 \div 2.5 = 175 \div 25 = 7$$

Order of Operations

When reviewing calculations consisting of more than one operation, the order in which the operations are performed affects the resulting answer. Consider $5 \times 2 + 7$. Performing multiplication then addition results in an answer of 17 because $(5 \times 2 = 10; 10 + 7 = 17)$. However, if the problem is written $5 \times (2 + 7)$, the order of operations dictates that the operation inside the parenthesis must be performed first. The resulting answer is 45 because $(2 + 7 = 9, \text{then } 5 \times 9 = 45)$.

The order in which operations should be performed is remembered using the acronym PEMDAS. PEMDAS stands for parenthesis, exponents, multiplication/division, addition/subtraction. Multiplication and division are performed in the same step, working from left to right with whichever comes first. Addition and subtraction are performed in the same step, working from left to right with whichever comes first.

Consider the following example: $8 \div 4 + 8(7 - 7)$. Performing the operation inside the parenthesis produces $8 \div 4 + 8(0)$ or $8 \div 4 + 8 \times 0$. There are no exponents, so multiplication and division are performed next from left to right resulting in: $2 + 8 \times 0$, then $2 + 0$. Finally, addition and subtraction are performed to obtain an answer of 2. Now consider the following example: $6 \times 3 + 3^2 - 6$. Parenthesis are not applicable. Exponents are evaluated first, which brings us to $6 \times 3 + 9 - 6$. Then multiplication/division forms $18 + 9 - 6$. At last, addition/subtraction leads to the final answer of 21.

Properties of Operations

As mentioned, properties of operations exist that make calculations easier and solve problems for missing values. The following table summarizes commonly used properties of real numbers.

Property	Addition	Multiplication
Commutative	$a + b = b + a$	$a \times b = b \times a$
Associative	$(a + b) + c = a + (b + c)$	$(a \times b) \times c = a \times (bc)$
Identity	$a + 0 = a; \ 0 + a = a$	$a \times 1 = a; \ 1 \times a = a$
Inverse	$a + (-a) = 0$	$a \times \dfrac{1}{a} = 1; \ a \neq 0$
Distributive	$a(b + c) = ab + ac$	

The commutative property of addition states that the order in which numbers are added does not change the sum. Similarly, the commutative property of multiplication states that the order in which numbers are multiplied does not change the product. The associative property of addition and multiplication state that the grouping of numbers being added or multiplied does not change the sum or product, respectively. The commutative and associative properties are useful for performing calculations. For example, $(47 + 25) + 3$ is equivalent to $(47 + 3) + 25$, which is easier to calculate.

The identity property of addition states that adding zero to any number does not change its value. The identity property of multiplication states that multiplying a number by 1 does not change its value. The inverse property of addition states that the sum of a number and its opposite equals zero. Opposites are numbers that are the same with different signs (ex. 5 and -5; $-\frac{1}{2}$ and $\frac{1}{2}$). The inverse property of multiplication states that the product of a number (other than 0) and its reciprocal equals 1. Reciprocal numbers have numerators and denominators that are inverted (ex. $\frac{2}{5}$ and $\frac{5}{2}$). Inverse properties are useful for canceling quantities to find missing values (see algebra content). For example, $a + 7 = 12$ is solved by adding the inverse of 7(-7) to both sides in order to isolate a.

The distributive property states that multiplying a sum (or difference) by a number produces the same result as multiplying each value in the sum (or difference) by the number and adding (or subtracting) the products. Consider the following scenario: You are buying three tickets for a baseball game. Each ticket costs $18. You are also charged a fee of $2 per ticket for purchasing the tickets online. The cost is calculated: $3 \times 18 + 3 \times 2$. Using the distributive property, the cost can also be calculated $3(18 + 2)$.

Arithmetic Word Problems

Word problems present the opportunity to relate mathematical concepts learned in the classroom into real-world situations. These types of problems are situations in which some parts of the problem are known and at least one part is unknown.

There are three types of instances in which something can be unknown: the starting point, the modification, or the final result can all be missing from the provided information.

- For an addition problem, the modification is the quantity of a new amount added to the starting point.

- For a subtraction problem, the modification is the quantity taken away from the starting point.

Keywords in the word problems can signal what type of operation needs to be used to solve the problem. Words such as *total, increased, combined*, and *more* indicate that addition is needed. Words such as *difference, decreased,* and *minus* indicate that subtraction is needed.

Consider the following addition equation: $3 + 7 = 10$.

The number 3 is the starting point, 7 is the modification, and 10 is the result from adding a new amount to the starting point. Different word problems can arise from this same equation, depending on which value is the unknown. For example, here are three problems:

- If a boy had three pencils and was given seven more, how many would he have in total?

- If a boy had three pencils and a girl gave him more so that he had ten in total, how many was he given?

- A boy was given seven pencils so that he had ten in total. How many did he start with?

All three problems involve the same equation, and determining which part of the equation is missing is the key to solving each word problem. The missing answers would be 10, 7, and 3, respectively.

When considering subtraction, the same three scenarios can occur. Let's consider the equation: $6 - 4 = 2$.

The number 6 is the starting point, 4 is the modification, and 2 is the new amount that is the result from taking away an amount from the starting point. Again, different types of word problems can arise from this equation. For example, here are three possible problems:

- If a girl had six quarters and two were taken away, how many would be left over?

- If a girl had six quarters, purchased a pencil, and had two quarters left over, how many did she pay with?

- If a girl paid for a pencil with four quarters and had two quarters left over, how many did she have to start with?

The three question types follow the structure of the addition word problems, and determining whether the starting point, the modification, or the final result is missing is the goal in solving the problem. The missing answers would be 2, 4, and 6, respectively.

The three addition problems and the three subtraction word problems can be solved by using a picture, a number line, or an algebraic equation. If an equation is used, a question mark can be utilized to represent the unknown quantity. For example, $6 - 4 =?$ can be written to show that the missing value is the result. Using equation form visually indicates what portion of the addition or subtraction problem is the missing value.

Similar instances can be seen in word problems involving multiplication and division. Key words within a multiplication problem involve *times, product, doubled*, and *tripled*. Key words within a division problem involve *split, quotient, divided, shared, groups*, and *half*. Like addition and subtraction, multiplication and division problems also have three different types of missing values.

Multiplication consists of a specific number of groups having the same size, the quantity of items within each group, and the total quantity within all groups. Therefore, each one of these amounts can be the missing value.

For example, consider the equation $5 \times 3 = 15$.

5 and 3 are interchangeable, so either amount can be the number of groups or the quantity of items within each group. 15 is the total number of items. Again, different types of word problems can arise from this equation. For example, here are three problems:

- If a classroom is serving 5 different types of apples for lunch and has three apples of each type, how many total apples are there to give to the students?

- If a classroom has 15 apples with 5 different types, how many of each type are there?

- If a classroom has 15 apples with 3 of each type, how many types are there to choose from?

Each question involves using the same equation, and it is imperative to decide which part of the equation is the missing value. The answers to the problems are 15, 3, and 5, respectively.

Similar to multiplication, division problems involve a total amount, a number of groups having the same size, and a number of items in each group. The difference between multiplication and division is that the starting point is the total amount, which then gets divided into equal quantities.

For example, consider the equation $15 \div 5 = 3$.

15 is the total number of items, which is being divided into 5 different groups. In order to do so, 3 items go into each group. Also, 5 and 3 are interchangeable, so the 15 items could be divided into 3 groups of 5 items each. Therefore, different types of word problems can arise from this equation depending on which value is unknown. For example, here are three types of problems:

- A boy needs 48 pieces of chalk. If there are 8 pieces in each box, how many boxes should he buy?

- A boy has 48 pieces of chalk. If each box has 6 pieces in it, how many boxes did he buy?

- A boy has partitioned all of his chalk into 8 piles, with 6 pieces in each pile. How many pieces does he have in total?

Each one of these questions involves the same equation, and the third question can easily utilize the multiplication equation $8 \times 6 = ?$ instead of division. The answers are 6, 8, and 48, respectively.

Word problems can appear daunting, but don't let the verbiage psyche you out. No matter the scenario or specifics, the key to answering them is to translate the words into a math problem. Always keep in mind what the question is asking and what operations could lead to that answer.

Word problems involving elapsed time, money, length, volume, and mass require:

- determining which operations (addition, subtraction, multiplication, and division) should be performed, and

- using and/or converting the proper unit for the scenario.

The following table lists key words which can be used to indicate the proper operation:

Addition	Sum, total, in all, combined, increase of, more than, added to
Subtraction	Difference, change, remaining, less than, decreased by
Multiplication	Product, times, twice, triple, each
Division	Quotient, goes into, per, evenly, divided by half, divided by third, split

Identifying and utilizing the proper units for the scenario requires knowing how to apply the conversion rates for money, length, volume, and mass. For example, given a scenario that requires subtracting 8 inches from $2\frac{1}{2}$ feet, both values should first be expressed in the same unit (they could be expressed $\frac{2}{3}$ft & $2\frac{1}{2}$ft, or 8in and 30in). The desired unit for the answer may also require converting back to another unit.

Consider the following scenario: A parking area along the river is only wide enough to fit one row of cars and is $\frac{1}{2}$ kilometers long. The average space needed per car is 5 meters. How many cars can be parked along the river? First, all measurements should be converted to similar units: $\frac{1}{2}$km = 500m. The operation(s) required should be identified. Because the problem asks for the number of cars, the total space should be divided by the space per car. Five-hundred meters divided by five meters per car yields a total of 100 cars. Written as an expression, the meters unit cancels and the cars unit is left: $\frac{500}{5m/car}$ is the same as $500m \times \frac{1\ car}{5m}$, which yields 100 cars.

When dealing with problems involving elapsed time, breaking the problem down into workable parts is helpful. For example, suppose the length of time between 1:15pm and 3:45pm must be determined. From 1:15pm to 2:00pm is 45 minutes (knowing there are 60 minutes in an hour). From 2:00pm to 3:00pm is 1 hour. From 3:00pm to 3:45pm is 45 minutes. The total elapsed time is 45 minutes plus 1 hour plus 45 minutes. This sum produces 1 hour and 90 minutes. 90 minutes is over an hour, so this is converted to1 hour (60 minutes) and 30 minutes. The total elapsed time can now be expressed as 2 hours and 30 minutes.

The following word problems highlight the most commonly tested question types.

<u>Working with Money</u>
Walter's Coffee Shop sells a variety of drinks and breakfast treats.

Price List		Costs	
Hot Coffee	$2.00	Hot Coffee	$0.25
Slow Drip Iced Coffee	$3.00	Slow Drip Iced Coffee	$0.75
Latte	$4.00	Latte	$1.00
Muffins	$2.00	Muffins	$1.00
Crepe	$4.00	Crepe	$2.00
Egg Sandwich	$5.00	Egg Sandwich	$3.00

Walter's utilities, rent, and labor costs him $500 per day. Today, Walter sold 200 hot coffees, 100 slow drip iced coffees, 50 lattes, 75 muffins, 45 crepes, and 60 egg sandwiches. What was Walter's total profit today?

To accurately answer this type of question, determine the total cost of making his drinks and treats, then determine how much revenue he earned from selling those products. After arriving at these two totals, the profit is measured by deducting the total cost from the total revenue.

Walter's costs for today:

Item	Quantity	Cost Per Unit	Total Cost
Hot Coffee	200	$0.25	$50
Slow-Drip Iced Coffee	100	$0.75	$75
Latte	50	$1.00	$50
Muffin	75	$1.00	$75
Crepe	45	$2.00	$90
Egg Sandwich	60	$3.00	$180
Utilities, rent, and labor			$500
Total Costs			$1,020

Walter's revenue for today:

Item	Quantity	Revenue Per Unit	Total Revenue
Hot Coffee	200	$2.00	$400
Slow-Drip Iced Coffee	100	$3.00	$300
Latte	50	$4.00	$200
Muffin	75	$2.00	$150
Crepe	45	$4.00	$180
Egg Sandwich	60	$5.00	$300
Total Revenue			$1,530

Walter's Profit = *Revenue − Costs* = $1,530 − $1,020 = $510

This strategy is applicable to other question types. For example, calculating salary after deductions, balancing a checkbook, and calculating a dinner bill are common word problems similar to business planning. Just remember to use the correct operations. When a balance is increased, use addition. When a balance is decreased, use subtraction. Common sense and organization are your greatest assets when answering word problems.

Converting Fractions, Decimals, and Percents

<u>Fractions</u>

A **fraction** is an equation that represents a part of a whole, but can also be used to present ratios or division problems. An example of a fraction is $\frac{x}{y}$. In this example, x is called the **numerator,** while y is the **denominator**. The numerator represents the number of parts, and the denominator is the total number of parts. They are separated by a line or slash. In simple fractions, the numerator and denominator can be nearly any integer. However, the denominator of a fraction can never be zero, because dividing by zero is a function which is undefined.

Imagine that an apple pie has been baked for a holiday party, and the full pie has eight slices. After the party, there are five slices left. How could the amount of the pie that remains be expressed as a fraction? The numerator is 5 since there are 5 pieces left, and the denominator is 8 since there were eight total slices in the whole pie. Thus, expressed as a fraction, the leftover pie totals $\frac{5}{8}$ of the original amount.

Fractions come in three different varieties: proper fractions, improper fractions, and mixed numbers. **Proper fractions** have a numerator less than the denominator, such as $\frac{3}{8}$, but **improper fractions** have a numerator greater than the denominator, such as $\frac{15}{8}$. **Mixed numbers** combine a whole number with a proper fraction, such as $3\frac{1}{2}$. Any mixed number can be written as an improper fraction by multiplying the integer by the denominator, adding the product to the value of the numerator, and dividing the sum by the original denominator. For example:

$$3\frac{1}{2} = \frac{3 \times 2 + 1}{2} = \frac{7}{2}$$

Whole numbers can also be converted into fractions by placing the whole number as the numerator and making the denominator 1. For example, $3 = \frac{3}{1}$.

One of the most fundamental concepts of fractions is their ability to be manipulated by multiplication or division. This is possible since $\frac{n}{n} = 1$ for any non-zero integer. As a result, multiplying or dividing by $\frac{n}{n}$ will not alter the original fraction since any number multiplied or divided by 1 doesn't change the value of that number. Fractions of the same value are known as equivalent fractions. For example, $\frac{2}{4}, \frac{4}{8}, \frac{50}{100}$, and $\frac{75}{150}$ are equivalent, as they all equal $\frac{1}{2}$.

Although many equivalent fractions exist, they are easier to compare and interpret when reduced or simplified. The numerator and denominator of a simple fraction will have no factors in common other than 1. When reducing or simplifying fractions, divide the numerator and denominator by the greatest common factor. A simple strategy is to divide the numerator and denominator by low numbers, like 2, 3, or 5 until arriving at a simple fraction, but the same thing could be achieved by determining the greatest common factor for both the numerator and denominator and dividing each by it. Using the first method is preferable when both the numerator and denominator are even, end in 5, or are obviously a multiple of another number. However, if no numbers seem to work, it will be necessary to factor the numerator and denominator to find the GCF.

Let's look at examples:

1) Simplify the fraction $\frac{6}{8}$:

Dividing the numerator and denominator by 2 results in $\frac{3}{4}$, which is a simple fraction.

2) Simplify the fraction $\frac{12}{36}$:

Dividing the numerator and denominator by 2 leaves $\frac{6}{18}$. This isn't a simple fraction, as both the numerator and denominator have factors in common. Diving each by 3 results in $\frac{2}{6}$, but this can be further simplified by dividing by 2 to get $\frac{1}{3}$. This is the simplest fraction, as the numerator is 1. In cases like this, multiple division operations can be avoided by determining the greatest common factor between the numerator and denominator.

3) Simplify the fraction $\frac{18}{54}$ by dividing by the greatest common factor:

First, determine the factors for the numerator and denominator. The factors of 18 are 1, 2, 3, 6, 9, and 18. The factors of 54 are 1, 2, 3, 6, 9, 18, 27, and 54. Thus, the greatest common factor is 18. Dividing $\frac{18}{54}$ by 18 leaves $\frac{1}{3}$, which is the simplest fraction. This method takes slightly more work, but it definitively arrives at the simplest fraction.

Of the four basic operations that can be performed on fractions, the one which involves the least amount of work is multiplication. To multiply two fractions, simply multiply the numerators, multiply the denominators, and place the products as a fraction. Whole numbers and mixed numbers can also be expressed as a fraction, as described above, to multiply with a fraction. Let's work through a couple of examples.

$$1)\ \frac{2}{5} \times \frac{3}{4} = \frac{6}{20} = \frac{3}{10}$$

$$2)\ \frac{4}{9} \times \frac{7}{11} = \frac{28}{99}$$

Dividing fractions is similar to multiplication with one key difference. To divide fractions, flip the numerator and denominator of the second fraction, and then proceed as if it were a multiplication problem:

$$1)\ \frac{7}{8} \div \frac{4}{5} = \frac{7}{8} \times \frac{5}{4} = \frac{35}{32}$$

$$2)\ \frac{5}{9} \div \frac{1}{3} = \frac{5}{9} \times \frac{3}{1} = \frac{15}{9} = \frac{5}{3}$$

Addition and subtraction require more steps than multiplication and division, as these operations require the fractions to have the same denominator, also called a common denominator. It is always possible to find a common denominator by multiplying the denominators. However, when the denominators are large numbers, this method is unwieldy, especially if the answer must be provided in its simplest form. Thus, it's beneficial to find the least common denominator of the fractions—the least common denominator is incidentally also the least common multiple.

Once equivalent fractions have been found with common denominators, simply add or subtract the numerators to arrive at the answer:

1) $\frac{1}{2} + \frac{3}{4} = \frac{2}{4} + \frac{3}{4} = \frac{5}{4}$

2) $\frac{3}{12} + \frac{11}{20} = \frac{15}{60} + \frac{33}{60} = \frac{48}{60} = \frac{4}{5}$

3) $\frac{7}{9} - \frac{4}{15} = \frac{35}{45} - \frac{12}{45} = \frac{23}{45}$

4) $\frac{5}{6} - \frac{7}{18} = \frac{15}{18} - \frac{7}{18} = \frac{8}{18} = \frac{4}{9}$

Decimals

The **decimal system** is a way of writing out numbers that uses ten different numerals: 0, 1, 2, 3, 4, 5, 6, 7, 8, and 9. This is also called a "base ten" or "base 10" system. Other bases are also used. For example, computers work with a base of 2. This means they only use the numerals 0 and 1.

The **decimal place** denotes how far to the right of the decimal point a numeral is. The first digit to the right of the decimal point is in the *tenths* place. The next is the **hundredths**. The third is the **thousandths**.

So, 3.142 has a 1 in the tenths place, a 4 in the hundredths place, and a 2 in the thousandths place.

The **decimal point** is a period used to separate the **ones** place from the **tenths** place when writing out a number as a decimal.

A **decimal number** is a number written out with a decimal point instead of as a fraction, for example, 1.25 instead of $\frac{5}{4}$. Depending on the situation, it can sometimes be easier to work with fractions and sometimes easier to work with decimal numbers.

A decimal number is **terminating** if it stops at some point. It is called **repeating** if it never stops, but repeats a pattern over and over. It is important to note that every rational number can be written as a terminating decimal or as a repeating decimal.

Addition with Decimals

To add decimal numbers, each number in columns needs to be lined up by the decimal point. For each number being added, the zeros to the right of the last number need to be filled in so that each of the numbers has the same number of places to the right of the decimal. Then, the columns can be added together. Here is an example of 2.45 + 1.3 + 8.891 written in column form:

$$2.450$$

$$1.300$$

$$+\,8.891$$

Zeros have been added in the columns so that each number has the same number of places to the right of the decimal.

Added together, the correct answer is 12.641:

$$2.450$$

$$1.300$$

$$\underline{+\ 8.891}$$

$$12.641$$

Subtraction with Decimals

Subtracting decimal numbers is the same process as adding decimals. Here is $7.89 - 4.235$ written in column form:

$$7.890$$

$$\underline{-\ 4.235}$$

$$3.655$$

A zero has been added in the column so that each number has the same number of places to the right of the decimal.

Multiplication with Decimals

Decimals can be multiplied as if there were no decimal points in the problem. For example, 0.5 x 1.25 can be rewritten and multiplied as 5 x 125, which equals 625.

The final answer will have the same number of decimal *points* as the total number of decimal *places* in the problem. The first number has one decimal place, and the second number has two decimal places. Therefore, the final answer will contain three decimal places:

$$0.5 \text{ x } 1.25 = 0.625$$

Division with Decimals

Dividing a decimal by a whole number entails using long division first by ignoring the decimal point. Then, the decimal point is moved the number of places given in the problem.

For example, $6.8 \div 4$ can be rewritten as $68 \div 4$, which is 17. There is one non-zero integer to the right of the decimal point, so the final solution would have one decimal place to the right of the solution. In this case, the solution is 1.7.

Dividing a decimal by another decimal requires changing the divisor to a whole number by moving its decimal point. The decimal place of the dividend should be moved by the same number of places as the divisor. Then, the problem is the same as dividing a decimal by a whole number.

For example, $5.72 \div 1.1$ has a divisor with one decimal point in the denominator. The expression can be rewritten as $57.2 \div 11$ by moving each number one decimal place to the right to eliminate the decimal. The long division can be completed as $572 \div 11$ with a result of 52. Since there is one non-zero integer to the right of the decimal point in the problem, the final solution is 5.2.

In another example, $8 \div 0.16$ has a divisor with two decimal points in the denominator. The expression can be rewritten as $800 \div 16$ by moving each number two decimal places to the right to eliminate the decimal in the divisor. The long division can be completed with a result of 50.

Percentages

Think of percentages as fractions with a denominator of 100. In fact, percentage means "per hundred." Problems often require converting numbers from percentages, fractions, and decimals.

The basic percent equation is the following:

$$\frac{is}{of} = \frac{\%}{100}$$

The placement of numbers in the equation depends on what the question asks.

Example 1: Find 40% of 80.

Basically, the problem is asking, "What is 40% of 80?" The 40% is the percent, and 80 is the number to find the percent "of." The equation is:

$$\frac{x}{80} = \frac{40}{100}$$

Solving the equation by cross-multiplication, the problem becomes $100x = 80(40)$. Solving for x gives the answer: $x = 32$.

Example 2: What percent of 100 is 20?

The 20 fills in the "is" portion, while 100 fills in the "of." The question asks for the percent, so that will be x, the unknown. The following equation is set up:

$$\frac{20}{100} = \frac{x}{100}$$

Cross-multiplying yields the equation $100x = 20(100)$. Solving for x gives the answer of 20%.

Example 3: 30% of what number is 30?

The following equation uses the clues and numbers in the problem:

$$\frac{30}{x} = \frac{30}{100}$$

Cross-multiplying results in the equation $30(100) = 30x$. Solving for x gives the answer $x = 100$.

Conversions

Decimals and Percentages

Since a percentage is based on "per hundred," decimals and percentages can be converted by multiplying or dividing by 100. Practically speaking, this always amounts to moving the decimal point two places to the right or left, depending on the conversion. To convert a percentage to a decimal, move the decimal point two places to the left and remove the % sign. To convert a decimal to a percentage, move the decimal point two places to the right and add a "%" sign.

Here are some examples:

65% = 0.65
0.33 = 33%
0.215 = 21.5%
99.99% = 0.9999
500% = 5.00
7.55 = 755%

Fractions and Percentages

Remember that a percentage is a number per one hundred. So a percentage can be converted to a fraction by making the number in the percentage the numerator and putting 100 as the denominator:

$$43\% = \frac{43}{100}$$

$$97\% = \frac{97}{100}$$

Note that the percent symbol (%) kind of looks like a 0, a 1, and another 0. So think of a percentage like 54% as 54 over 100.

To convert a fraction to a percent, follow the same logic. If the fraction happens to have 100 in the denominator, you're in luck. Just take the numerator and add a percent symbol:

$$\frac{28}{100} = 28\%$$

Otherwise, divide the numerator by the denominator to get a decimal:

$$\frac{9}{12} = 0.75$$

Then convert the decimal to a percentage:

$$0.75 = 75\%$$

Another option is to make the denominator equal to 100. Be sure to multiply the numerator by the same number as the denominator. For example:

$$\frac{3}{20} \times \frac{5}{5} = \frac{15}{100}$$

$$\frac{15}{100} = 15\%$$

Changing Fractions to Decimals

To change a fraction into a decimal, divide the denominator into the numerator until there are no remainders. There may be repeating decimals, so rounding is often acceptable. A straight line above the repeating portion denotes that the decimal repeats.

Example: Express 4/5 as a decimal.

Set up the division problem.

$$5\overline{)4}$$

5 does not go into 4, so place the decimal and add a zero.

$$5\overline{)4.0}$$

5 goes into 40 eight times. There is no remainder.

$$\begin{array}{r} 0.8 \\ 5\overline{)4.0} \\ -4.0 \\ \hline 0 \end{array}$$

The solution is 0.8.

Example: Express 33 1/3 as a decimal.

Since the whole portion of the number is known, set it aside to calculate the decimal from the fraction portion.

Set up the division problem.

$$3\overline{)1}$$

3 does not go into 1, so place the decimal and add zeros. 3 goes into 10 three times.

$$\begin{array}{r} 0.333 \\ 3\overline{)1.000} \end{array}$$

This will repeat with a remainder of 3, so place a line over the 3 denotes the repetition.

$$\begin{array}{r} 0.333 \\ 3\overline{)1.000} \\ -9 \\ \hline 10 \\ -9 \\ \hline 10 \end{array}$$

The solution is $0.\overline{3}$

Changing Decimals to Fractions
To change decimals to fractions, place the decimal portion of the number, the numerator, over the respective place value, the denominator, then reduce, if possible.

Example: Express 0.25 as a fraction.

This is read as twenty-five hundredths, so put 25 over 100. Then reduce to find the solution.

$$\frac{25}{100} = \frac{1}{4}$$

Example: Express 0.455 as a fraction

This is read as four hundred fifty-five thousandths, so put 455 over 1000. Then reduce to find the solution.

$$\frac{455}{1000} = \frac{91}{200}$$

There are two types of problems that commonly involve percentages. The first is to calculate some percentage of a given quantity, where you convert the percentage to a decimal, and multiply the quantity by that decimal. Secondly, you are given a quantity and told it is a fixed percent of an unknown quantity. In this case, convert to a decimal, then divide the given quantity by that decimal.

Example: What is 30% of 760?

Convert the percent into a useable number. "Of" means to multiply.

$$30\% = 0.30$$

Set up the problem based on the givens, and solve.

$$0.30 \times 760 = 228$$

Example: 8.4 is 20% of what number?

Convert the percent into a useable number.

$$20\% = 0.20$$

The given number is a percent of the answer needed, so divide the given number by this decimal rather than multiplying it.

$$\frac{8.4}{0.20} = 42$$

Ratios, Rates, and Proportions

Ratios
Ratios are used to show the relationship between two quantities. The ratio of oranges to apples in the grocery store may be 3 to 2. That means that for every 3 oranges, there are 2 apples. This comparison can be expanded to represent the actual number of oranges and apples. Another example may be the number of boys to girls in a math class. If the ration of boys to girls is given as 2 to 5, that means there are 2 boys to every 5 girls in the class. Ratios can also be compared if the units in each ratio are the same. The ratio of boys to girls in the math class can be compared to the ratio of boys to girls in a science class by stating which ratio is higher and which is lower.

Rates are used to compare two quantities with different units. **Unit rates** are the simplest form of rate. With unit rates, the denominator in the comparison of two units is one. For example, if someone can type at a rate of 1000 words in 5 minutes, then his or her unit rate for typing is $\frac{1000}{5} = 200$ words in one minute or 200 words per minute. Any rate can be converted into a unit rate by dividing to make the denominator one. 1000 words in 5 minutes has been converted into the unit rate of 200 words per minute.

<u>Unit Rate</u>

Unit rate word problems will ask you to calculate the rate or quantity of something in a different value. For example, a problem might tell you that a car drove a certain number of miles in a certain number of minutes and then ask how many miles per hour the car was traveling. These questions involve solving proportions. Consider the following examples:

1. Alexandra made $96 during the first 3 hours of her shift as a temporary worker at a law office. She will continue to earn money at this rate until she finishes in 5 more hours. How much does Alexandra make per hour? How much will Alexandra have made at the end of the day?

This problem can be solved in two ways. The first is to set up a proportion because the rate of pay is constant. The second is to determine her hourly rate, multiply the 5 hours by that rate, and then add the $96. We'll solve it using both methods.

To set up a proportion, put the money already earned over the hours already worked on one side of an equation. The other side has x over 8 hours (the total hours worked in the day). It looks like this: $\frac{96}{3} = \frac{x}{8}$. Now, cross-multiply to get $768 = 3x$. To get x, we divide by 3, which leaves us with $x = 256$. Thus, Alexandra will make $256 at the end of the day. To calculate her hourly rate, we need to divide the total by 8, giving us $32 per hour.

Alternatively, we could figure out the hourly rate by dividing $96 by 3 hours to get $32 per hour. Now we can figure out her total pay by multiplying $32 per hour by 8 hours, which comes out to $256.

2. Jonathan is reading a novel. So far, he has read 215 of the 335 total pages. It takes Jonathan 25 minutes to read 10 pages, and the rate is constant. How long does it take Jonathan to read one page? How much longer will it take him to finish the novel? Express the answer in time.

To calculate how long it takes Jonathan to read one page, we need to divide the 25 minutes by 10 pages to determine the page per minute rate. Thus, it takes 2.5 minutes to read one page.

Jonathan must read 120 more pages to complete the novel. (This is calculated by subtracting the pages already read from the total.) Now, we need to multiply his rate per page by the number of pages. Thus, $120 \times 2.5 = 300$. Expressed in time, 300 minutes is equal to 5 hours.

3. At a hotel, $\frac{4}{5}$ of the 120 rooms are booked for Saturday. On Sunday, $\frac{3}{4}$ of the 120 rooms are booked. On which day are more of the rooms booked, and by how much more?

The first step is to calculate the number of rooms booked for each day. We can do this by multiplying the fraction of the rooms booked by the total number of rooms.

$$\text{Saturday:} \frac{4}{5} \times 120 = \frac{4}{5} \times \frac{120}{1} = \frac{480}{5} = 96 \text{ rooms}$$

$$\text{Sunday:} \frac{3}{4} \times 120 = \frac{3}{4} \times \frac{120}{1} = \frac{360}{4} = 90 \text{ rooms}$$

Thus, more rooms were booked on Saturday by 6 rooms.

4. In a veterinary hospital, the veterinarian-to-pet ratio is 1:9. The ratio is always constant. If there are 45 pets in the hospital, how many veterinarians are currently in the veterinary hospital?

We can set up a proportion to solve for the number of veterinarians: $\frac{1}{9} = \frac{x}{45}$

After cross-multiplying, we're left with $9x = 45$, which works out to 5 veterinarians.

Alternatively, as there are always 9 times as many pets as veterinarians, we can divide the number of pets (45) by 9. This also arrives at the correct answer of 5 veterinarians.

5. At a general practice law firm, 30% of the lawyers work solely on tort cases. If 9 lawyers work solely on tort cases, how many lawyers work at the firm?

We need to solve for the total number of lawyers working at the firm, which we will represent with x. We know that 9 lawyers work solely on torts cases, and they make up 30% of the total lawyers at the firm. Thus, 30% multiplied by the total, x, will equal 9. Written as equation, this is: $30\% \times x = 9$.

It's easier to deal with the equation if we convert the percentage to a decimal, leaving us with $0.3x = 9$. Thus, $x = \frac{9}{0.3} = 30$ lawyers working at the firm.

6. Xavier was hospitalized with pneumonia. He was originally given 35mg of antibiotics. Later, after his condition continued to worsen, Xavier's dosage was increased to 60mg. What was the percent increase of the antibiotics? Round the percentage to the nearest tenth.

For questions asking for the increase in percentage, we need to calculate the difference, divide by the original amount, and multiply by 100. Written as an equation, the formula is:

$$\frac{new\ quanity - old\ quantity}{old\ quantity} \times 100$$

This formula also works if you are asked for the percentage decrease.

Here, the question tells us that the dosage increased from 35mg to 60mg, so we can plug those numbers into the formula to find the percentage increase.

$$\frac{60 - 35}{35} \times 100 = \frac{25}{35} \times 100 = .7142 \times 100 = 71.4\%$$

Using Ratio Reasoning to Convert Rates
Ratios and rates can be used together to convert rates into different units. For example, if someone is driving 50 kilometers per hour, that rate can be converted into miles per hour by using a ratio known as the **conversion factor**. Since the given value contains kilometers and the final answer needs to be in miles, the ratio relating miles to kilometers needs to be used. There are 0.62 miles in 1 kilometer. This, written as a ratio and in fraction form, is $\frac{0.62\ miles}{1\ km}$. To convert 50km/hour into miles per hour, the following conversion needs to be set up:

$$\frac{50\ km}{hour} \times \frac{0.62\ miles}{1\ km} = 31\ miles\ per\ hour$$

Solving Problems Involving Scale Factors

The ratio between two similar geometric figures is called the **scale factor**. For example, a problem may depict two similar triangles, A and B. The scale factor from the smaller triangle A to the larger triangle B is given as 2 because the length of the corresponding side of the larger triangle, 16, is twice the corresponding side on the smaller triangle, 8. This scale factor can also be used to find the value of a missing side, x, in triangle A. Since the scale factor from the smaller triangle (A) to larger one (B) is 2, the larger corresponding side in triangle B (given as 25), can be divided by 2 to find the missing side in A (x = 12.5). The scale factor can also be represented in the equation $2A = B$ because two times the lengths of A gives the corresponding lengths of B. This is the idea behind similar triangles.

Proportional Relationships

Much like a scale factor can be written using an equation like $2A = B$, a **relationship** is represented by the equation $Y = kX$. X and Y are **proportional** because as values of X increase, the values of Y also increase. A relationship that is **inversely proportional** can be represented by the equation $Y = \frac{k}{x}$, where the value of Y decreases as the value of x increases and vice versa.

Proportional reasoning can be used to solve problems involving ratios, percentages, and averages. Ratios can be used in setting up proportions and solving them to find unknowns. For example, if a student completes an average of 10 pages of math homework in 3 nights, how long would it take the student to complete 22 pages? Both ratios can be written as fractions. The second ratio would contain the unknown.

The following proportion represents this problem, where x is the unknown number of nights:

$$\frac{10 \; pages}{3 \; nights} = \frac{22 \; pages}{x \; nights}$$

Solving this proportion entails cross-multiplying and results in the following equation:

$$10x = 22 \times 3$$

Simplifying and solving for x results in the exact solution: $x = 6.6 \; nights$. The result would be rounded up to 7 because the homework would actually be completed on the 7th night.

The following problem uses ratios involving percentages:

If 20% of the class is girls and 30 students are in the class, how many girls are in the class?

To set up this problem, it is helpful to use the common proportion:

$$\frac{\%}{100} = \frac{is}{of}$$

Within the proportion, % is the percentage of girls, 100 is the total percentage of the class, *is* is the number of girls, and *of* is the total number of students in the class. Most percentage problems can be written using this language. To solve this problem, the proportion should be set up as $\frac{20}{100} = \frac{x}{30}$, and then solved for x. Cross-multiplying results in the equation $20 \times 30 = 100x$, which results in the solution $x = 6$. There are 6 girls in the class.

Problems involving volume, length, and other units can also be solved using ratios. For example, a problem may ask for the volume of a cone to be found that has a radius, $r = 7m$ and a height, $h = 16m$. Referring to the formulas provided on the test, the volume of a cone is given as: $V = \pi r^2 \frac{h}{3}$, where r is the radius, and h is the height. Plugging $r = 7$ and $h = 16$ into the formula, the following is obtained:

$$V = \pi(7^2)\frac{16}{3}$$

Therefore, volume of the cone is found to be approximately 821m³. Sometimes, answers in different units are sought. If this problem wanted the answer in liters, 821m³ would need to be converted. Using the equivalence statement 1m³ = 1000L, the following ratio would be used to solve for liters:

$$821\text{m}^3 \times \frac{1000L}{1m^3}$$

Cubic meters in the numerator and denominator cancel each other out, and the answer is converted to 821,000 liters, or 8.21×10^5 L.

Other conversions can also be made between different given and final units. If the temperature in a pool is 30°C, what is the temperature of the pool in degrees Fahrenheit? To convert these units, an equation is used relating Celsius to Fahrenheit. The following equation is used:

$$T_{°F} = 1.8T_{°C} + 32$$

Plugging in the given temperature and solving the equation for T yields the result:

$$T_{°F} = 1.8(30) + 32 = 86°F$$

Both units in the metric system and U.S. customary system are widely used.

Here are some more examples of how to solve for proportions:

1) $\frac{75\%}{90\%} = \frac{25\%}{x}$

To solve for x, the fractions must be cross multiplied:

$$(75\%x = 90\% \times 25\%)$$

To make things easier, let's convert the percentages to decimals:

$$(0.9 \times 0.25 = 0.225 = 0.75x)$$

To get rid of x's co-efficient, each side must be divided by that same coefficient to get the answer $x = 0.3$. The question could ask for the answer as a percentage or fraction in lowest terms, which are 30% and $\frac{3}{10}$, respectively.

2) $\frac{x}{12} = \frac{30}{96}$

Cross-multiply: $96x = 30 \times 12$

Multiply: $96x = 360$

Divide: $x = 360 \div 96$

Answer: $x = 3.75$

3) $\frac{0.5}{3} = \frac{x}{6}$

Cross-multiply: $3x = 0.5 \times 6$

Multiply: $3x = 3$

Divide: $x = 3 \div 3$

Answer: $x = 1$

You may have noticed there's a faster way to arrive at the answer. If there is an obvious operation being performed on the proportion, the same operation can be used on the other side of the proportion to solve for x. For example, in the first practice problem, 75% became 25% when divided by 3, and upon doing the same to 90%, the correct answer of 30% would have been found with much less legwork. However, these questions aren't always so intuitive, so it's a good idea to work through the steps, even if the answer seems apparent from the outset.

Solving Ratio and Percent Problems
Questions dealing with percentages can be difficult when they are phrased as word problems. These word problems almost always come in three varieties. The first type will ask to find what percentage of some number will equal another number. The second asks to determine what number is some percentage of another given number. The third will ask what number another number is a given percentage of.

One of the most important parts of correctly answering percentage word problems is to identify the numerator and the denominator. This fraction can then be converted into a percentage, as described above.

The following word problem shows how to make this conversion:

A department store carries several different types of footwear. The store is currently selling 8 athletic shoes, 7 dress shoes, and 5 sandals. What percentage of the store's footwear are sandals?

First, calculate what serves as the 'whole', as this will be the denominator. How many total pieces of footwear does the store sell? The store sells 20 different types (8 athletic + 7 dress + 5 sandals).

Second, what footwear type is the question specifically asking about? Sandals. Thus, 5 is the numerator.

Third, the resultant fraction must be expressed as a percentage. The first two steps indicate that $\frac{5}{20}$ of the footwear pieces are sandals. This fraction must now be converted into a percentage:

$$\frac{5}{20} \times \frac{5}{5} = \frac{25}{100} = 25\%$$

Estimation

Estimation is finding a value that is close to a solution but is not the exact answer. For example, if there are values in the thousands to be multiplied, then each value can be estimated to the nearest thousand and the calculation performed. This value provides an approximate solution that can be determined very quickly.

Rounding is the process of either bumping a number up or down, based on a specified place value. First, the place value is specified. Then, the digit to its right is looked at. For example, if rounding to the nearest hundreds place, the digit in the tens place is used. If it is a 0, 1, 2, 3, or 4, the digit being rounded to is left alone. If it is a 5, 6, 7, 8 or 9, the digit being rounded to is increased by one. All other digits before the decimal point are then changed to zeros, and the digits in decimal places are dropped. If a decimal place is being rounded to, all subsequent digits are just dropped. For example, if 845,231.45 was to be rounded to the nearest thousands place, the answer would be 845,000. The 5 would remain the same due to the 2 in the hundreds place. Also, if 4.567 was to be rounded to the nearest tenths place, the answer would be 4.6. The 5 increased to 6 due to the 6 in the hundredths place, and the rest of the decimal is dropped.

Sometimes when performing operations such as multiplying numbers, the result can be estimated by **rounding.** For example, to estimate the value of 11.2×2.01, each number can be rounded to the nearest integer. This will yield a result of 22.

Rounding numbers helps with estimation because it changes the given number to a simpler, although less accurate, number than the exact given number. Rounding allows for easier calculations, which estimate the results of using the exact given number. The accuracy of the estimate and ease of use depends on the place value to which the number is rounded. Rounding numbers consists of:

- determining what place value the number is being rounded to

- examining the digit to the right of the desired place value to decide whether to round up or keep the digit, and

- replacing all digits to the right of the desired place value with zeros.

To round 746,311 to the nearest ten thousand, the digit in the ten thousands place should be located first. In this case, this digit is 4 (7<u>4</u>6,311). Then, the digit to its right is examined. If this digit is 5 or greater, the number will be rounded up by increasing the digit in the desired place by one. If the digit to the right of the place value being rounded is 4 or less, the number will be kept the same. For the given example, the digit being examined is a 6, which means that the number will be rounded up by increasing the digit to the left by one. Therefore, the digit 4 is changed to a 5. Finally, to write the rounded number, any digits to the left of the place value being rounded remain the same and any to its right are replaced with zeros. For the given example, rounding 746,311 to the nearest ten thousand will produce 750,000. To round 746,311 to the nearest hundred, the digit to the right of the three in the hundreds place is examined to determine whether to round up or keep the same number. In this case, that digit is a 1, so the number will be kept the same and any digits to its right will be replaced with zeros. The resulting rounded number is 746,300.

Rounding place values to the right of the decimal follows the same procedure, but digits being replaced by zeros can simply be dropped. To round 3.752891 to the nearest thousandth, the desired place value is located (3.75$\underline{2}$891) and the digit to the right is examined. In this case, the digit 8 indicates that the number will be rounded up, and the 2 in the thousandths place will increase to a 3. Rounding up and replacing the digits to the right of the thousandths place with zeros produces 3.753000, which is equivalent to 3.753. Therefore, the zeros are not necessary and the rounded number should be written as 3.753.

When rounding up, if the digit to be increased is a 9, the digit to its left is increased by 1 and the digit in the desired place value is changed to a zero. For example, the number 1,598 rounded to the nearest ten is 1,600. Another example shows the number 43.72961 rounded to the nearest thousandth is 43.730 or 43.73.

Mental math should always be considered as problems are worked through, and the ability to work through problems in one's head helps save time. If a problem is simple enough, such as $15 + 3 = 18$, it should be completed mentally. The ability to do this will increase once addition and subtraction in higher place values are grasped. Also, mental math is important in multiplication and division. The times tables multiplying all numbers from 1 to 12 should be memorized. This will allow for division within those numbers to be memorized as well. For example, we should know easily that $121 \div 11 = 11$ because it should be memorized that $11 \times 11 = 121$. Here is the multiplication table to be memorized:

x	1	2	3	4	5	6	7	8	9	10	11	12	13	14	15
1	1	2	3	4	5	6	7	8	9	10	11	12	13	14	15
2	2	4	6	8	10	12	14	16	18	20	22	24	26	28	30
3	3	6	9	12	15	18	21	24	27	30	33	36	39	42	45
4	4	8	12	16	20	24	28	32	36	40	44	48	52	56	60
5	5	10	15	20	25	30	35	40	45	50	55	60	65	70	75
6	6	12	18	24	30	36	42	48	54	60	66	72	78	84	90
7	7	14	21	28	35	42	49	56	63	70	77	84	91	98	105
8	8	16	24	32	40	48	56	64	72	80	88	96	104	112	120
9	9	18	27	36	45	54	63	72	81	90	99	108	117	126	135
10	10	20	30	40	50	60	70	80	90	100	110	120	130	140	150
11	11	22	33	44	55	66	77	88	99	110	121	132	143	154	165
12	12	24	36	48	60	72	84	96	108	120	132	144	156	168	180
13	13	26	39	52	65	78	91	104	117	130	143	156	169	182	195
14	14	28	42	56	70	84	98	112	126	140	154	168	182	196	210
15	15	30	45	60	75	90	105	120	135	150	165	180	195	210	225

The values in yellow along the diagonal of the table consist of perfect squares. A **perfect square** is a number that represents a product of two equal integers.

Patterns

Patterns within a sequence can come in 2 distinct forms: the items (shapes, numbers, etc.) either repeat in a constant order, or the items change from one step to another in some consistent way. The **core** is the smallest unit, or number of items, that repeats in a repeating pattern. For example, the pattern

○○▲○○▲○... has a core that is ○○▲. Knowing only the core, the pattern can be extended. Knowing the number of steps in the core allows the identification of an item in each step without drawing/writing the entire pattern out. For example, suppose the tenth item in the previous pattern must be determined. Because the core consists of three items (○○▲), the core repeats in multiples of 3. In other words, steps 3, 6, 9, 12, etc. will be ▲ completing the core with the core starting over on the next step. For the above example, the 9th step will be ▲ and the 10th will be ○.

The most common patterns in which each item changes from one step to the next are arithmetic and geometric sequences. An **arithmetic sequence** is one in which the items increase or decrease by a constant difference. In other words, the same thing is added or subtracted to each item or step to produce the next. To determine if a sequence is arithmetic, determine what must be added or subtracted to step one to produce step two. Then, check if the same thing is added/subtracted to step two to produce step three. The same thing must be added/subtracted to step three to produce step four, and so on. Consider the pattern 13, 10, 7, 4 . . . To get from step one (13) to step two (10) by adding or subtracting requires subtracting by 3. The next step is checking if subtracting 3 from step two (10) will produce step three (7), and subtracting 3 from step three (7) will produce step four (4). In this case, the pattern holds true. Therefore, this is an arithmetic sequence in which each step is produced by subtracting 3 from the previous step. To extend the sequence, 3 is subtracted from the last step to produce the next. The next three numbers in the sequence are 1, -2, -5.

A **geometric sequence** is one in which each step is produced by multiplying or dividing the previous step by the same number. To determine if a sequence is geometric, decide what step one must be multiplied or divided by to produce step two. Then check if multiplying or dividing step two by the same number produces step three, and so on. Consider the pattern 2, 8, 32, 128 . . . To get from step one (2) to step two (8) requires multiplication by 4. The next step determines if multiplying step two (8) by 4 produces step three (32), and multiplying step three (32) by 4 produces step four (128). In this case, the pattern holds true. Therefore, this is a geometric sequence in which each step is produced by multiplying the previous step by 4. To extend the sequence, the last step is multiplied by 4 and repeated. The next three numbers in the sequence are 512; 2,048; 8,192.

Although arithmetic and geometric sequences typically use numbers, these sequences can also be represented by shapes. For example, an arithmetic sequence could consist of shapes with three sides, four sides, and five sides (add one side to the previous step to produce the next). A geometric sequence could consist of eight blocks, four blocks, and two blocks (each step is produced by dividing the number of blocks in the previous step by 2).

Frequencies

In mathematics, **frequencies** refer to how often an event occurs or the number of times a particular quantity appears in a given series. To find the number of times a specific value appears, frequency tables are used to record the occurrences, which can then be summed. To construct a frequency table, one simply inputs the values into a tabular format with a column denoting each value, typically in ascending order, with a second column to tally up the number of occurrences for each value, and a third column to give a numerical frequency based on the number of tallies.

A **frequency distribution** communicates the number of outcomes of a given value or number in a data set. When displayed as a bar graph or histogram, it can visually indicate the spread and distribution of the data. A histogram resembling a bell curve approximates a normal distribution. A frequency distribution can also be displayed as a **stem-and-leaf plot**, which arranges data in numerical order and displays values similar to a tally chart with the stem being a range within the set and the leaf indicating the exact value. (For example, stems are whole numbers and leaves are tenths.)

	Movie Ratings
4	7
5	2 6 9
6	1 4 6 8 8
7	0 3 5 9
8	1 3 5 6 8 8 9
9	0 0 1 3 4 6 6 9

Key   6 | 1 represents 61

Stem	Leaf
2	0 2 3 6 8 8 9
3	2 6 7 7
4	7 9
5	4 6 9

This plot provides more detail about individual data points and allows for easy identification of the median, as well as any repeated values in the set.

Data that isn't described using numbers is known as **categorical data.** For example, age is numerical data but hair color is categorical data. Categorical data can also be summarized using two-way frequency tables. A **two-way frequency table** counts the relationship between two sets of categorical data. There are rows and columns for each category, and each cell represents frequency information that shows the actual data count between each combination. For example, below is a two-way frequency table showing the gender and breed of cats in an animal shelter:

	Domestic Shorthair	Persian	Domestic Longhair	Total
Male	12	2	7	21
Female	8	4	5	17
Total	20	6	12	38

Entries in the middle of the table are known as the **joint frequencies**. For example, the number of female cats that are Persian is 4, which is a joint frequency. The totals are the **marginal frequencies**. For example, the total number of male cats is 21, which is a marginal frequency. If the frequencies are changed into percentages based on totals, the table is known as a **two-way relative frequency table**. Percentages can be calculated using the table total, the row totals, or the column totals. Two-way frequency tables can help in making conclusions about the data.

Algebra

Algebraic Expressions and Equations

An **algebraic expression** is a statement about an unknown quantity expressed in mathematical symbols. A variable is used to represent the unknown quantity, usually denoted by a letter. An **equation** is a statement in which two expressions (at least one containing a variable) are equal to each other. An algebraic expression can be thought of as a mathematical phrase and an equation can be thought of as a mathematical sentence.

Algebraic expressions and equations both contain numbers, variables, and mathematical operations. The following are examples of algebraic expressions: $5x + 3$, $7xy - 8(x^2 + y)$, and $\sqrt{a^2 + b^2}$. An expression can be simplified or evaluated for given values of variables. The following are examples of equations: $2x + 3 = 7$, $a^2 + b^2 = c^2$, and $2x + 5 = 3x - 2$. An equation contains two sides separated by an equal sign. Equations can be solved to determine the value(s) of the variable for which the statement is true.

Adding and Subtracting Linear Algebraic Expressions

An algebraic expression is simplified by combining like terms. A **term** is a number, variable, or product of a number, and variables separated by addition and subtraction. For the algebraic expression $3x^2 - 4x + 5 - 5x^2 + x - 3$, the terms are $3x^2$, -4x, 5, $-5x^2$, x, and -3. **Like terms** have the same variables raised to the same powers (exponents). The like terms for the previous example are $3x^2$ and $-5x^2$, -4x and x, 5 and -3. To combine like terms, the coefficients (numerical factor of the term including sign) are added and the variables and their powers are kept the same. Note that if a coefficient is not written, it is an implied coefficient of 1 ($x = 1x$). The previous example will simplify to $-2x^2 - 3x + 2$.

When adding or subtracting algebraic expressions, each expression is written in parenthesis. The negative sign is distributed when necessary, and like terms are combined. Consider the following: add $2a + 5b - 2$ to $a - 2b + 8c - 4$. The sum is set as follows:

$$(a - 2b + 8c - 4) + (2a + 5b - 2)$$

In front of each set of parentheses is an implied positive one, which, when distributed, does not change any of the terms. Therefore, the parentheses are dropped and like terms are combined:

$$a - 2b + 8c - 4 + 2a + 5b - 2 = 3a + 3b + 8c - 6$$

Consider the following problem: Subtract $2a + 5b - 2$ from $a - 2b + 8c - 4$. The difference is set as follows:

$$(a - 2b + 8c - 4) - (2a + 5b - 2)$$

The implied one in front of the first set of parentheses will not change those four terms. However, distributing the implied -1 in front of the second set of parentheses will change the sign of each of those three terms:

$$a - 2b + 8c - 4 - 2a - 5b + 2$$

Combining like terms yields the simplified expression:

$$-a - 7b + 8c - 2$$

Distributive Property

The **distributive property** states that multiplying a sum (or difference) by a number produces the same result as multiplying each value in the sum (or difference) by the number and adding (or subtracting) the products. Using mathematical symbols, the distributive property states $a(b + c) = ab + ac$. The expression $4(3 + 2)$ is simplified using the order of operations. Simplifying inside the parenthesis first produces 4×5, which equals 20. The expression $4(3 + 2)$ can also be simplified using the distributive property:

$$4(3 + 2) = 4 \times 3 + 4 \times 2$$

$$12 + 8 = 20$$

Consider the following example: $4(3x - 2)$. The expression cannot be simplified inside the parenthesis because $3x$ and -2 are not like terms, and therefore cannot be combined. However, the expression can be simplified by using the distributive property and multiplying each term inside of the parenthesis by the term outside of the parenthesis: $12x - 8$. The resulting equivalent expression contains no like terms, so it cannot be further simplified.

Consider the expression:

$$(3x + 2y + 1) - (5x - 3) + 2(3y + 4)$$

Again, there are no like terms, but the distributive property is used to simplify the expression. Note there is an implied one in front of the first set of parentheses and an implied -1 in front of the second set of parentheses. Distributing the one, -1, and 2 produces:

$$1(3x) + 1(2y) + 1(1) - 1(5x) - 1(-3) + 2(3y) + 2(4)$$

$$3x + 2y + 1 - 5x + 3 + 6y + 8$$

This expression contains like terms that are combined to produce the simplified expression:

$$-2x + 8y + 12$$

Algebraic expressions are tested to be equivalent by choosing values for the variables and evaluating both expressions. For example, $4(3x - 2)$ and $12x - 8$ are tested by substituting 3 for the variable x and calculating to determine if equivalent values result.

Simple Expressions for Given Values

As mentioned, an **algebraic expression** is a statement written in mathematical symbols, typically including one or more unknown values represented by variables. For example, the expression $2x +$

3 states that an unknown number (x) is multiplied by 2 and added to 3. If given a value for the unknown number, or variable, the value of the expression can be determined. For example, if the value of the variable x is 4, the value of the expression 4 is multiplied by 2, and 3 is added. This results in a value of 11 for the expression.

When given an algebraic expression and values for the variable(s), the expression is evaluated to determine its numerical value. To evaluate the expression, the given values for the variables are substituted (or replaced) and the expression is simplified using the order of operations. Parenthesis should be used when substituting. Consider the following: Evaluate $a - 2b + ab$ for $a = 3$ and $b = -1$. To evaluate, any variable a is replaced with 3 and any variable b with -1, producing:

$$3 - 2(-1) + (3)(-1)$$

Next, the order of operations is used to calculate the value of the expression, which is 2.

Parts of Expressions

Algebraic expressions consist of variables, numbers, and operations. A term of an expression is any combination of numbers and/or variables, and terms are separated by addition and subtraction. For example, the expression $5x^2 - 3xy + 4 - 2$ consists of 4 terms: $5x^2$, -3xy, 4y, and -2. Note that each term includes its given sign (+ or −). The **variable** part of a term is a letter that represents an unknown quantity. The **coefficient** of a term is the number by which the variable is multiplied. For the term $4y$, the variable is y and the coefficient is 4. Terms are identified by the power (or exponent) of its variable.

A number without a variable is referred to as a **constant**. If the variable is to the first power (x^1 or simply x), it is referred to as a **linear term**. A term with a variable to the second power (x^2) is **quadratic** and a term to the third power (x^3) is **cubic**. Consider the expression $x^3 + 3x - 1$. The constant is -1. The linear term is $3x$. There is no quadratic term. The cubic term is x^3.

An algebraic expression can also be classified by how many terms exist in the expression. Any like terms should be combined before classifying. A **monomial** is an expression consisting of only one term. Examples of monomials are: 17, 2x, and $-5ab^2$. A **binomial** is an expression consisting of two terms separated by addition or subtraction. Examples include:

$$2x - 4 \text{ and } -3y^2 + 2y$$

A **trinomial** consists of 3 terms. For example, $5x^2 - 2x + 1$ is a trinomial.

Verbal Statements and Algebraic Expressions

An algebraic expression is a statement about unknown quantities expressed in mathematical symbols. The statement *five times a number added to forty* is expressed as $5x + 40$. An equation is a statement in which two expressions (with at least one containing a variable) are equal to one another. The statement *five times a number added to forty is equal to ten* is expressed as $5x + 40 = 10$.

Real-world scenarios can also be expressed mathematically. Suppose a job pays its employees $300 per week and $40 for each sale made. The weekly pay is represented by the expression $40x + 300$ where x is the number of sales made during the week.

Consider the following scenario: Bob had $20 and Tom had $4. After selling 4 ice cream cones to Bob, Tom has as much money as Bob. The cost of an ice cream cone is an unknown quantity and can be represented by a variable (x). The amount of money Bob has after his purchase is four times the cost of an ice cream cone subtracted from his original \$20 → $20 - 4x$. The amount of money Tom has after his sale is four times the cost of an ice cream cone added to his original \$4 → $4x + 4$. After the sale, the amount of money that Bob and Tom have are equal → $20 - 4x = 4x + 4$.

When expressing a verbal or written statement mathematically, it is vital to understand words or phrases that can be represented with symbols. The following are examples:

Symbol	Phrase
+	Added to; increased by; sum of; more than
−	Decreased by; difference between; less than; take away
×	Multiplied by; 3(4,5...) times as large; product of
÷	Divided by; quotient of; half (third, etc.) of
=	Is; the same as; results in; as much as; equal to
x,t,n, etc.	A number; unknown quantity; value of; variable

Use of Formulas

Formulas are mathematical expressions that define the value of one quantity, given the value of one or more different quantities. Formulas look like equations because they contain variables, numbers, operators, and an equal sign. All formulas are equations but not all equations are formulas. A formula must have more than one variable. For example, $2x + 7 = y$ is an equation and a formula (it relates the unknown quantities x and y). However, $2x + 7 = 3$ is an equation but not a formula (it only expresses the value of the unknown quantity x).

Formulas are typically written with one variable alone (or isolated) on one side of the equal sign. This variable can be thought of as the **subject** in that the formula is stating the value of the subject in terms of the relationship between the other variables. Consider the distance formula: $distance = rate \times time$ or $d = rt$. The value of the subject variable d (distance) is the product of the variable r and t (rate and time). Given the rate and time, the distance traveled can easily be determined by substituting the values into the formula and evaluating.

The formula $P = 2l + 2w$ expresses how to calculate the perimeter of a rectangle (P) given its length (l) and width (w). To find the perimeter of a rectangle with a length of 3ft and a width of 2ft, these values are substituted into the formula for l and w:

$$P = 2(3ft) + 2(2ft)$$

Following the order of operations, the perimeter is determined to be 10ft. When working with formulas such as these, including units is an important step.

Given a formula expressed in terms of one variable, the formula can be manipulated to express the relationship in terms of any other variable. In other words, the formula can be rearranged to change which variable is the subject. To solve for a variable of interest by manipulating a formula, the equation may be solved as if all other variables were numbers. The same steps for solving are followed, leaving operations in terms of the variables instead of calculating numerical values. For the formula $P = 2l + 2w$, the perimeter is the subject expressed in terms of the length and width. To write a formula to calculate the width of a rectangle, given its length and perimeter, the previous formula relating the three variables is solved for the variable w. If P and l were numerical values, this is a two-step linear equation solved by subtraction and division. To solve the equation $P = 2l + 2w$ for w, $2l$ is first subtracted from both sides:

$$P - 2l = 2w$$

Then both sides are divided by 2:

$$\frac{P - 2l}{2} = w$$

Dependent and Independent Variables

A **variable** represents an unknown quantity and, in the case of a formula, a specific relationship exists between the variables. Within a given scenario, variables are the quantities that are changing. If two variables exist, one is dependent and one is independent. The value of one variable depends on the other variable. If a scenario describes distance traveled and time traveled at a given speed, distance is dependent and time is independent. The distance traveled depends on the time spent traveling. If a scenario describes the cost of a cab ride and the distance traveled, the cost is dependent and the distance is independent. The cost of a cab ride depends on the distance travelled. Formulas often contain more than two variables and are typically written with the dependent variable alone on one side of the equation. This lone variable is the subject of the statement. If a formula contains three or more variables, one variable is dependent and the rest are independent. The values of all independent variables are needed to determine the value of the dependent variable.

The formula $P = 2l + 2w$ expresses the dependent variable P in terms of the independent variables, l and w. The perimeter of a rectangle depends on its length and width. The formula $d = rt$ ($distance = rate \times time$) expresses the dependent variable d in terms of the independent variables, r and t. The distance traveled depends on the rate (or speed) and the time traveled.

Multistep One-Variable Linear Equations and Inequalities

Linear equations and linear inequalities are both comparisons of two algebraic expressions. However, unlike equations in which the expressions are equal, linear inequalities compare expressions that may be unequal. Linear equations typically have one value for the variable that makes the statement true. Linear inequalities generally have an infinite number of values that make the statement true.

When solving a linear equation, the desired result requires determining a numerical value for the unknown variable. If given a linear equation involving addition, subtraction, multiplication, or division, working backwards isolates the variable. Addition and subtraction are inverse operations, as are multiplication and division. Therefore, they can be used to cancel each other out.

The first steps to solving linear equations are distributing, if necessary, and combining any like terms on the same side of the equation. Sides of an equation are separated by an equal sign. Next, the equation is manipulated to show the variable on one side. Whatever is done to one side of the equation must be done to the other side of the equation to remain equal. Inverse operations are then used to isolate the variable and undo the order of operations backwards. Addition and subtraction are undone, then multiplication and division are undone.

For example, solve $4(t - 2) + 2t - 4 = 2(9 - 2t)$

Distributing: $4t - 8 + 2t - 4 = 18 - 4t$

Combining like terms: $6t - 12 = 18 - 4t$

Adding $4t$ to each side to move the variable: $10t - 12 = 18$

Adding 12 to each side to isolate the variable: $10t = 30$

Dividing each side by 10 to isolate the variable: $t = 3$

The answer can be checked by substituting the value for the variable into the original equation, ensuring that both sides calculate to be equal.

Linear inequalities express the relationship between unequal values. More specifically, they describe in what way the values are unequal. A value can be greater than (>), less than (<), greater than or equal to (≥), or less than or equal to (≤) another value. $5x + 40 > 65$ is read as *five times a number added to forty is greater than sixty-five.*

When solving a linear inequality, the solution is the set of all numbers that make the statement true. The inequality $x + 2 \geq 6$ has a solution set of 4 and every number greater than 4 (4.01; 5; 12; 107; etc.). Adding 2 to 4 or any number greater than 4 results in a value that is greater than or equal to 6. Therefore, $x \geq 4$ is the solution set.

To algebraically solve a linear inequality, follow the same steps as those for solving a linear equation. The inequality symbol stays the same for all operations *except* when multiplying or dividing by a negative number. If multiplying or dividing by a negative number while solving an inequality, the relationship reverses (the sign flips). In other words, > switches to < and vice versa. Multiplying or dividing by a positive number does not change the relationship, so the sign stays the same. An example is shown below.

Solve $-2x - 8 \leq 22$

Add 8 to both sides: $-2x \leq 30$

Divide both sides by -2: $x \geq -15$

Solutions of a linear equation or a linear inequality are the values of the variable that make a statement true. In the case of a linear equation, the solution set (list of all possible solutions) typically consists of a single numerical value. To find the solution, the equation is solved by isolating the variable. For example, solving the equation $3x - 7 = -13$ produces the solution $x = -2$. The only value for x that produces a true statement is -2. This can be checked by substituting -2 into the original equation to check that both sides are equal. In this case, $3(-2) - 7 = -13 \rightarrow -13 = -13$; therefore, -2 is a solution.

Although linear equations generally have one solution, this is not always the case. If there is no value for the variable that makes the statement true, there is no solution to the equation. Consider the equation:

$$x + 3 = x - 1$$

There is no value for *x* in which adding 3 to the value produces the same result as subtracting one from the value. Conversely, if any value for the variable makes a true statement, the equation has an infinite number of solutions. Consider the equation:

$$3x + 6 = 3(x + 2)$$

Any number substituted for *x* will result in a true statement (both sides of the equation are equal).

By manipulating equations like the two above, the variable of the equation will cancel out completely. If the remaining constants express a true statement (ex. $6 = 6$), then all real numbers are solutions to the equation. If the constants left express a false statement (ex. $3 = -1$), then no solution exists for the equation.

Solving a linear inequality requires all values that make the statement true to be determined. For example, solving $3x - 7 \geq -13$ produces the solution $x \geq -2$. This means that -2 and any number greater than -2 produces a true statement. Solution sets for linear inequalities will often be displayed using a number line. If a value is included in the set ($\geq$ or $\leq$), a shaded dot is placed on that value and an arrow extending in the direction of the solutions. For a variable > or $\geq$ a number, the arrow will point right on a number line, the direction where the numbers increase. If a variable is < or $\leq$ a number, the arrow will point left on a number line, which is the direction where the numbers decrease. If the value is not included in the set (> or <), an open (unshaded) circle on that value is used with an arrow in the appropriate direction.

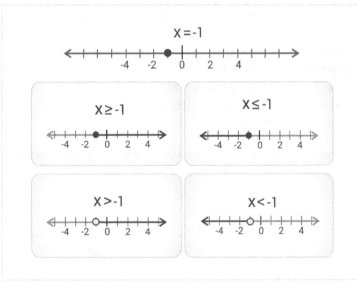

Similar to linear equations, a linear inequality may have a solution set consisting of all real numbers, or can contain no solution. When solved algebraically, a linear inequality in which the variable cancels out and results in a true statement (ex. $7 \geq 2$) has a solution set of all real numbers. A linear inequality in which the variable cancels out and results in a false statement (ex. $7 \leq 2$) has no solution.

Linear Relationships

Linear relationships describe the way two quantities change with respect to each other. The relationship is defined as linear because a line is produced if all the sets of corresponding values are graphed on a coordinate grid. When expressing the linear relationship as an equation, the equation is often written in the form $y = mx + b$ (**slope-intercept form**) where m and b are numerical values and x and y are variables (for example, $y = 5x + 10$). Given a linear equation and the value of either variable (x or y), the value of the other variable can be determined.

Suppose a teacher is grading a test containing 20 questions with 5 points given for each correct answer, adding a curve of 10 points to each test. This linear relationship can be expressed as the equation $y = 5x + 10$ where x represents the number of correct answers and y represents the test score. To determine the score of a test with a given number of correct answers, the number of correct answers is substituted into the equation for x and evaluated. For example, for 10 correct answers, 10 is substituted for x:

$$y = 5(10) + 10 \rightarrow y = 60$$

Therefore, 10 correct answers will result in a score of 60. The number of correct answers needed to obtain a certain score can also be determined. To determine the number of correct answers needed to score a 90, 90 is substituted for y in the equation (y represents the test score) and solved: $90 = 5x + 10 \rightarrow 80 = 5x \rightarrow 16 = x$. Therefore, 16 correct answers are needed to score a 90.

Linear relationships may be represented by a table of 2 corresponding values. Certain tables may determine the relationship between the values and predict other corresponding sets. Consider the table below, which displays the money in a checking account that charges a monthly fee:

Month	0	1	2	3	4
Balance	$210	$195	$180	$165	$150

An examination of the values reveals that the account loses $15 every month (the month increases by one and the balance decreases by 15). This information can be used to predict future values. To determine what the value will be in month 6, the pattern can be continued, and it can be concluded that the balance will be $120. To determine which month the balance will be $0, $210 is divided by $15 (since the balance decreases $15 every month), resulting in month 14.

Similar to a table, a graph can display corresponding values of a linear relationship.

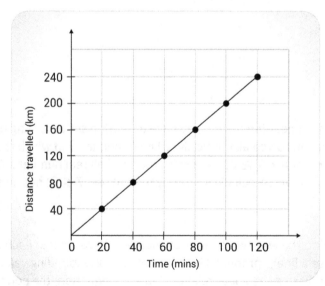

The graph above represents the relationship between distance traveled and time. To find the distance traveled in 80 minutes, the mark for 80 minutes is located at the bottom of the graph. By following this mark directly up on the graph, the corresponding point for 80 minutes is directly across from the 160 kilometer mark. This information indicates that the distance travelled in 80 minutes is 160 kilometers. To predict information not displayed on the graph, the way in which the variables change with respect to one another is determined. In this case, distance increases by 40 kilometers as time increases by 20 minutes. This information can be used to continue the data in the graph or convert the values to a table.

Conjectures, Predictions, or Generalizations Based on Patterns

An arithmetic or geometric sequence can be written as a formula and used to determine unknown steps without writing out the entire sequence. (Note that a similar process for repeating patterns is covered in the previous section.) An arithmetic sequence progresses by a **common difference**. To determine the common difference, any ste

p is subtracted by the step that precedes it. In the sequence 4, 9, 14, 19 . . . the common difference, or d, is 5. By expressing each step as a_1, a_2, a_3, etc., a formula can be written to represent the sequence. a_1 is the first step. To produce step two, step 1 (a_1) is added to the common difference (d):

$$a_2 = a_1 + d$$

To produce step three, the common difference (d) is added twice to a_1:

$$a_3 = a_1 + 2d$$

To produce step four, the common difference (d) is added three times to a_1:

$$a_4 = a_1 + 3d$$

Following this pattern allows a general rule for arithmetic sequences to be written. For any term of the sequence (a_n), the first step (a_1) is added to the product of the common difference (d) and one less than the step of the term ($n - 1$):

$$a_n = a_1 + (n - 1)d$$

Suppose the 8th term (a_8) is to be found in the previous sequence. By knowing the first step (a_1) is 4 and the common difference (d) is 5, the formula can be used:

$$a_n = a_1 + (n - 1)d$$

$$a_8 = 4 + (7)5$$

$$a_8 = 39$$

In a geometric sequence, each step is produced by multiplying or dividing the previous step by the same number. The **common ratio**, or (r), can be determined by dividing any step by the previous step. In the sequence 1, 3, 9, 27 . . . the common ratio (r) is 3 ($\frac{3}{1} = 3$ or $\frac{9}{3} = 3$ or $\frac{27}{9} = 3$). Each successive step can be expressed as a product of the first step (a_1) and the common ratio (r) to some power. For example:

$$a_2 = a_1 \times r$$

$$a_3 = a_1 \times r \times r$$

103

$$a_3 = a_1 \times r^2$$

$$a_4 = a_1 \times r \times r \times r$$

$$a_4 = a\_1 \times r^3$$

Following this pattern, a general rule for geometric sequences can be written. For any term of the sequence (a_n), the first step (a_1) is multiplied by the common ratio (r) raised to the power one less than the step of the term ($n - 1$):

$$a_n = a_1 \times r^{(n-1)}$$

Suppose for the previous sequence, the 7th term (a_7) is to be found. Knowing the first step (a_1) is one, and the common ratio (r) is 3, the formula can be used:

$$a_n = a_1 \times r^{(n-1)}$$

$$a_7 = (1) \times 3^6$$

$$a_7 = 729$$

Corresponding Terms of Two Numerical Patterns

When given two numerical patterns, the corresponding terms should be examined to determine if a relationship exists between them. Corresponding terms between patterns are the pairs of numbers that appear in the same step of the two sequences. Consider the following patterns 1, 2, 3, 4 . . . and 3, 6, 9, 12 . . . The corresponding terms are: 1 and 3; 2 and 6; 3 and 9; and 4 and 12. To identify the relationship, each pair of corresponding terms is examined and the possibilities of performing an operation (+, −, ×, ÷) to the term from the first sequence to produce the corresponding term in the second sequence are determined. In this case:

$1 + 2 = 3$	or	$1 \times 3 = 3$
$2 + 4 = 6$	or	$2 \times 3 = 6$
$3 + 6 = 9$	or	$3 \times 3 = 9$
$4 + 8 = 12$	or	$4 \times 3 = 12$

The consistent pattern is that the number from the first sequence multiplied by 3 equals its corresponding term in the second sequence. By assigning each sequence a label (input and output) or variable (x and y), the relationship can be written as an equation. If the first sequence represents the inputs, or x, and the second sequence represents the outputs, or y, the relationship can be expressed as: $y = 3x$.

Consider the following sets of numbers:

a	2	4	6	8
b	6	8	10	12

To write a rule for the relationship between the values for *a* and the values for *b*, the corresponding terms (2 and 6; 4 and 8; 6 and 10; 8 and 12) are examined. The possibilities for producing *b* from *a* are:

$$2 + 4 = 6 \qquad \text{or} \qquad 2 \times 3 = 6$$

$$4 + 4 = 8 \qquad \text{or} \qquad 4 \times 2 = 8$$

$$6 + 4 = 10$$

$$8 + 4 = 12 \qquad \text{or} \qquad 8 \times 1.5 = 12$$

The consistent pattern is that adding 4 to the value of *a* produces the value of *b*. The relationship can be written as the equation $a + 4 = b$.

Geometry & Measurement

Geometry is part of mathematics. It deals with shapes and their properties. It is also similar to measurement and number operations. The basis of geometry involves being able to label and describe shapes and their properties. That knowledge will lead to working with formulas such as area, perimeter, and volume. This knowledge will help to solve word problems involving shapes.

Flat or two-dimensional shapes include circles, triangles, hexagons, and rectangles, among others. Three-dimensional solid shapes, such as spheres and cubes, are also used in geometry. A shape can be classified based on whether it is open like the letter U or closed like the letter O. Further classifications involve counting the number of sides and vertices (corners) on the shapes. This will help you tell the difference between shapes.

Polygons can be drawn by sketching a fixed number of line segments that meet to create a closed shape. In addition, **triangles** can be drawn by sketching a closed space using only three-line segments. **Quadrilaterals** are closed shapes with four-line segments. Note that a triangle has three vertices, and a quadrilateral has four vertices.

To draw **circles**, one curved line segment must be drawn that has only one endpoint. This creates a closed shape. Given such direction, every point on the line would be the same distance away from its center. The **radius** of a circle goes from an endpoint on the center of the circle to an endpoint on the circle. The **diameter** is the line segment created by placing an endpoint on the circle, drawing through the radius, and placing the other endpoint on the circle. A compass can be used to draw circles of a more precise size and shape.

Points, Lines, Line Segments, and Rays

The basic unit of geometry is a point. A **point** represents an exact location on a plane, or flat surface. The position of a point is indicated with a dot and usually named with a single uppercase letter, such as point

A or point *T*. A point is a place, not a thing, and therefore has no dimensions or size. A set of points that lies on the same line is called **collinear**. A set of points that lies on the same plane is called **coplanar**.

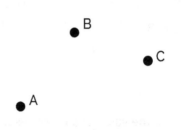

The image above displays point *A*, point *B*, and point *C*.

A **line** is as series of points that extends in both directions without ending. It consists of an infinite number of points and is drawn with arrows on both ends to indicate it extends infinitely. Lines can be named by two points on the line or with a single, cursive, lower case letter. The two lines below could be named line *AB* or line *BA* or $\overleftrightarrow{AB}$ or $\overleftrightarrow{BA}$; and line *m*.

Two lines are considered parallel to each other if, while extending infinitely, they will never intersect (or meet). **Parallel** lines point in the same direction and are always the same distance apart. Two lines are considered **perpendicular** if they intersect to form right angles. Right angles are 90°. Typically, a small box is drawn at the intersection point to indicate the right angle.

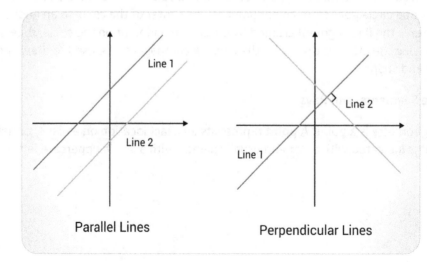

Line 1 is parallel to line 2 in the left image and is written as line 1 || line 2. Line 1 is perpendicular to line 2 in the right image and is written as line 1 ⊥ line 2.

A **ray** has a specific starting point and extends in one direction without ending. The endpoint of a ray is its starting point. Rays are named using the endpoint first, and any other point on the ray. The following ray can be named ray *AB* and written $\overrightarrow{AB}$.

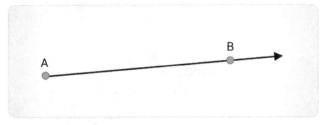

A **line segment** has specific starting and ending points. A line segment consists of two endpoints and all the points in between. Line segments are named by the two endpoints. The example below is named segment *KL* or segment *LK*, written $\overline{KL}$ or $\overline{LK}$.

Angles

An **angle** consists of two rays that have a common endpoint. This common endpoint is called the **vertex** of the angle. The two rays can be called sides of the angle. The angle below has a vertex at point *B* and the sides consist of ray *BA* and ray *BC*. An angle can be named in three ways:

1. Using the vertex and a point from each side, with the vertex letter in the middle.
2. Using only the vertex. This can only be used if it is the only angle with that vertex.
3. Using a number that is written inside the angle.

The angle below can be written ∠*ABC* (read angle *ABC*), ∠*CBA*, ∠*B*, or ∠1.

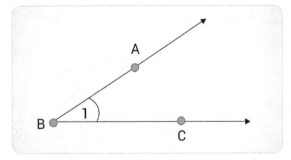

An angle divides a **plane**, or flat surface, into three parts: the angle itself, the interior (inside) of the angle, and the exterior (outside) of the angle. The figure below shows point *M* on the interior of the angle and point *N* on the exterior of the angle.

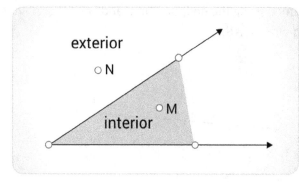

Angles can be measured in units called degrees, with the symbol °. The degree measure of an angle is between 0° and 180°, is a measure of rotation, and can be obtained by using a **protractor**, which is the tool pictured below.

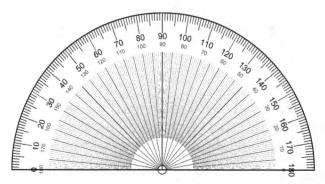

To use a protractor to measure an angle, the vertex, or corner, of the angle goes in the midpoint of the protractor, in that small circle along the bottom straight edge. Then one line of the angle is lined up along the bottom edge, toward the 0° indicator. Then, the degrees are read wherever the other line of the angle crosses.

In the example below, it can be seen that the angle measures about 30°:

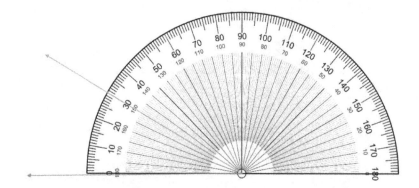

A straight angle (or simply a line) measures exactly 180°. A right angle's sides meet at the vertex to create a square corner. A right angle measures exactly 90° and is typically indicated by a box drawn in the interior of the angle. An acute angle has an interior that is narrower than a right angle. The measure of an acute angle is any value less than 90° and greater than 0°. For example, 89.9°, 47°, 12°, and 1°. An obtuse angle has an interior that is wider than a right angle. The measure of an obtuse angle is any value greater than 90° but less than 180°. For example, 90.1°, 110°, 150°, and 179.9°. Any two angles that sum up to 90 degrees are known as **complementary angles**.

- Acute angles: Less than 90°
- Obtuse angles: Greater than 90°
- Right angles: 90°
- Straight angles: 180°

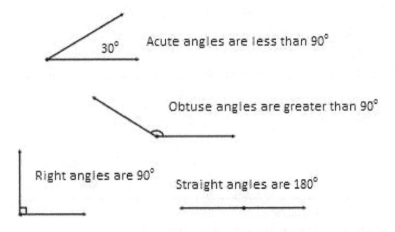

Determining the Measure of an Unknown Angle

To determine angle measures for adjacent angles, angles that share a common side and vertex, at least one of the angles must be known. Other information that is necessary to determine such measures include that there are 90° in a right angle, and there are 180° in a straight line. Therefore, if two adjacent angles form a right angle, they will add up to 90°, and if two adjacent angles form a straight line, they add up to 180°.

If the measurement of one of the adjacent angles is known, the other can be found by subtracting the known angle from the total number of degrees.

For example, given the following situation, if angle *a* measures 55⁰, find the measure of unknown angle *b*:

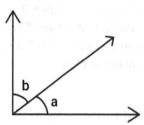

To solve this simply subtract the known angle measure from 90⁰.

$$90° - 55° = 35°$$

The measure of *b* = 35°.

Given the following situation, if angle 1 measures 45⁰, find the measure of the unknown angle 2:

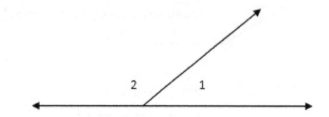

To solve this, simply subtract the known angle measure from 180⁰.

$$180° - 45° = 135°$$

The measure of angle 2 = 135°.

In the case that more than two angles are given, use the same method of subtracting the known angles from the total measure.

For example, given the following situation, if angle *y* = 40⁰, and angle *z* = 25⁰, find unknown angle *x*.

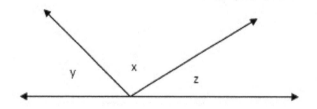

Subtract the known angles from 180⁰.

$$180° - 40° - 25° = 115°$$

The measure of angle *x* = 115⁰.

Classifying Two-Dimensional Figures

The vocabulary regarding many two-dimensional shapes is an important part classifying these figures. Flat or two-dimensional shapes include circles, triangles, hexagons, and rectangles, among others. A shape can be classified based on whether it is open like the letter U or closed like the letter O. Further classifications involve counting the number of sides and vertices (corners) on the shapes. This can help to tell the difference between shapes.

A **polygon** is a closed geometric figure in a plane (flat surface) consisting of at least 3 sides formed by line segments.

Polygons can be either convex or concave. A polygon that has interior angles all measuring less than 180° is **convex**. A **concave** polygon has one or more interior angles measuring greater than 180°. Examples are shown below.

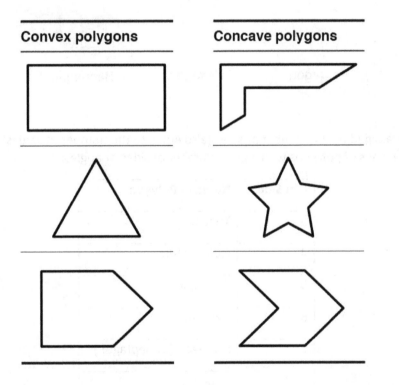

Polygons are often defined as two-dimensional shapes. Common two-dimensional shapes include circles, triangles, squares, rectangles, pentagons, and hexagons. A **circle** is a two-dimensional shape without sides, so it is not a polygon.

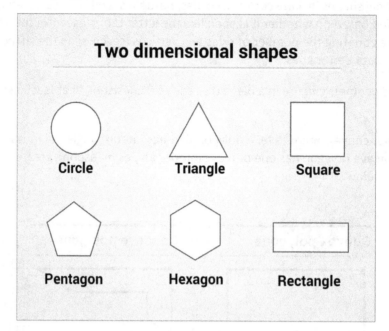

Polygons can be classified by the number of sides (also equal to the number of angles) they have. The following are the names of polygons with a given number of sides or angles:

# of Sides	Name of Polygon
3	Triangle
4	Quadrilateral
5	Pentagon
6	Hexagon
7	Septagon (or heptagon)
8	Octagon
9	Nonagon
10	Decagon

Equiangular polygons are polygons in which the measure of every interior angle is the same. The sides of equilateral polygons are always the same length. If a polygon is both equiangular and equilateral, the polygon is defined as a **regular polygon**. Examples are shown below:

Many four-sided figures can also be identified using properties of angles and lines. A **quadrilateral** is a closed shape with four sides. A **parallelogram** is a specific type of quadrilateral that has two sets of parallel lines that have the same length. A **trapezoid** is a quadrilateral having only one set of parallel sides. A **rectangle** is a parallelogram that has four right angles. A **rhombus** is a parallelogram with two acute angles, two obtuse angles, and four equal sides. The acute angles are of equal measure, and the obtuse angles are of equal measure. Finally, a **square** is a rhombus consisting of four right angles. It is important to note that some of these shapes share common attributes. For instance, all four-sided shapes are quadrilaterals.

All squares are rectangles, but not all rectangles are squares.

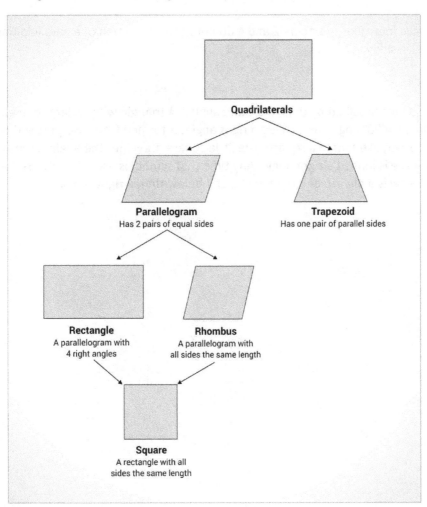

Consider the following:

Which of the following are not rectangles?

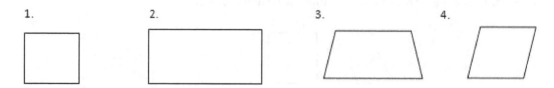

Even though all have four sides and some are even parallel sides, numbers 3 and 4 do not fit the definition of a rectangle.

Which of the following are not parallelograms?

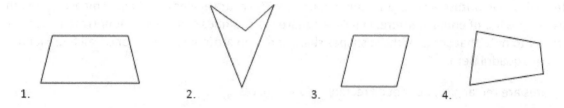

Even though all have four sides, numbers 2 and 4 do not fit the definition of a parallelogram because they have no parallel sides.

Classifying Triangles

Triangles can be further classified by their sides and angles. A triangle with its largest angle measuring 90° is a **right triangle**. A 90° angle, also called a **right angle,** is formed from two perpendicular lines. It looks like a hard corner, like that in a square. The little square draw into the angle is the symbol used to denote that that angle is indeed a right angle. Any time that symbol is used, it denotes the measure of the angle is 90°. Below is a picture of a right angle, and below that, a right triangle.

A right angle:

Here is a right triangle, which is a triangle that contains a right angle:

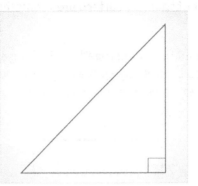

A triangle with the largest angle less than 90° is an acute triangle. A triangle with the largest angle greater than 90° is an obtuse triangle.

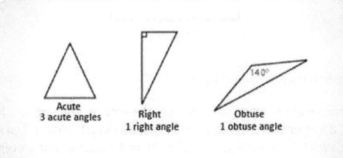

A triangle consisting of two equal sides and two equal angles is an isosceles triangle. A triangle with three equal sides and three equal angles is an equilateral triangle. A triangle with no equal sides or angles is a scalene triangle.

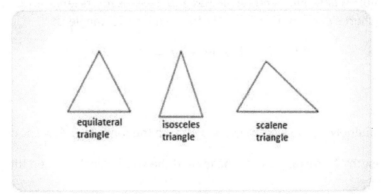

Area and Perimeter

Perimeter is the measurement of a distance around something or the sum of all sides of a polygon. Think of perimeter as the length of the boundary, like a fence. In contrast, **area** is the space occupied by a defined enclosure, like a field enclosed by a fence.

When thinking about perimeter, think about walking around the outside of something. When thinking about area, think about the amount of space or **surface area** something takes up.

Square
The perimeter of a square is measured by adding together all of the sides. Since a square has four equal sides, its perimeter can be calculated by multiplying the length of one side by 4. Thus, the formula is $P = 4 \times s$, where s equals one side. For example, the following square has side lengths of 5 meters:

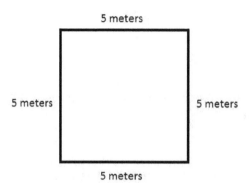

The perimeter is 20 meters because 4 times 5 is 20.

The area of a square is the length of a side squared, and the area of a rectangle is length multiplied by the width. For example, if the length of the square is 7 centimeters, then the area is 49 square centimeters. The formula for this example is $A = s^2 = 7^2 = 49$ square centimeters. An example is if the rectangle has a length of 6 inches and a width of 7 inches, then the area is 42 square inches:

$$A = lw = 6(7) = 42 \text{ square inches}$$

Rectangle
Like a square, a rectangle's perimeter is measured by adding together all of the sides. But as the sides are unequal, the formula is different. A rectangle has equal values for its lengths (long sides) and equal values for its widths (short sides), so the perimeter formula for a rectangle is:

$$P = l + l + w + w = 2l + 2w$$

l equals length
w equals width

The area is found by multiplying the length by the width, so the formula is $A = l \times w$.

For example, if the length of a rectangle is 10 inches and the width 8 inches, then the perimeter is 36 inches because:

$$P = 2l + 2w = 2(10) + 2(8)$$

$$20 + 16 = 36 \text{ inches}$$

Triangle

A triangle's perimeter is measured by adding together the three sides, so the formula is $P = a + b + c$, where a, b, and c are the values of the three sides. The area is the product of one-half the base and height so the formula is:

$$A = \frac{1}{2} \times b \times h$$

It can be simplified to:

$$A = \frac{bh}{2}$$

The base is the bottom of the triangle, and the height is the distance from the base to the peak. If a problem asks to calculate the area of a triangle, it will provide the base and height.

For example, if the base of the triangle is 2 feet and the height 4 feet, then the area is 4 square feet. The following equation shows the formula used to calculate the area of the triangle:

$$A = \frac{1}{2}bh = \frac{1}{2}(2)(4) = 4 \text{ square feet}$$

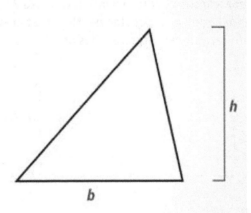

Circle

A circle's perimeter—also known as its circumference—is measured by multiplying the diameter by π.

Diameter is the straight line measured from one end to the direct opposite end of the circle.

π is referred to as pi and is equal to 3.14 (with rounding).

So the formula is $\pi \times d$.

This is sometimes expressed by the formula $C = 2 \times \pi \times r$, where r is the radius of the circle. These formulas are equivalent, as the radius equals half of the diameter.

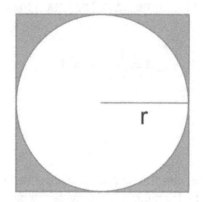

The area of a circle is calculated through the formula $A = \pi \times r^2$. The test will indicate either to leave the answer with π attached or to calculate to the nearest decimal place, which means multiplying by 3.14 for π.

Surface Area and Volume of Geometric Shapes

The area of a two-dimensional figure refers to the number of square units needed to cover the interior region of the figure. This concept is similar to wallpaper covering the flat surface of a wall. For example, if a rectangle has an area of 8 square inches (written 8 in^2), it will take 8 squares, each with sides one inch in length, to cover the interior region of the rectangle. Note that area is measured in square units such as: square feet or ft^2; square yards or yd^2; square miles or mi^2.

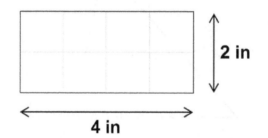

The surface area of a three-dimensional figure refers to the number of square units needed to cover the entire surface of the figure. This concept is similar to using wrapping paper to completely cover the outside of a box. For example, if a triangular pyramid has a surface area of 17 square inches (written 17in^2), it will take 17 squares, each with sides one inch in length, to cover the entire surface of the pyramid. Surface area is also measured in square units.

Many three-dimensional figures (solid figures) can be represented by nets consisting of rectangles and triangles. The surface area of such solids can be determined by adding the areas of each of its faces and bases. Finding the surface area using this method requires calculating the areas of rectangles and triangles. To find the area (A) of a rectangle, the length (l) is multiplied by the width (w) $\rightarrow A = l \times w$. The area of a rectangle with a length of 8cm and a width of 4cm is calculated:

$$A = (8cm) \times (4cm) \rightarrow A = 32cm^2$$

To calculate the area (A) of a triangle, the product of $\frac{1}{2}$, the base (b), and the height (h) is found →

$$A = \frac{1}{2} \times b \times h$$

Note that the height of a triangle is measured from the base to the vertex opposite of it forming a right angle with the base. The area of a triangle with a base of 11cm and a height of 6cm is calculated:

$$A = \frac{1}{2} \times (11cm) \times (6cm) \rightarrow A = 33cm^2$$

Consider the following triangular prism, which is represented by a net consisting of two triangles and three rectangles.

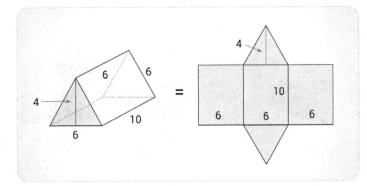

The surface area of the prism can be determined by adding the areas of each of its faces and bases. The surface area (SA) = area of triangle + area of triangle + area of rectangle + area of rectangle + area of rectangle.

$$SA = \left(\frac{1}{2} \times b \times h\right) + \left(\frac{1}{2} \times b \times h\right) + (l \times w) + (l \times w) + (l \times w)$$

$$SA = \left(\frac{1}{2} \times 6 \times 4\right) + \left(\frac{1}{2} \times 6 \times 4\right) + (6 \times 10) + (6 \times 10) + (6 \times 10)$$

$$SA = (12) + (12) + (60) + (60) + (60)$$

$$SA = 204 \ square \ units$$

Volume is the measurement of how much space an object occupies, like how much space is in the cube. Volume is useful in determining the space within a certain three-dimensional object. Volume can be calculated for a cube, rectangular prism, cylinder, pyramid, cone, and sphere. By knowing specific dimensions of the objects, the volume of the object is computed with these figures. The units for the volumes of solids can include cubic centimeters, cubic meters, cubic inches, and cubic feet.

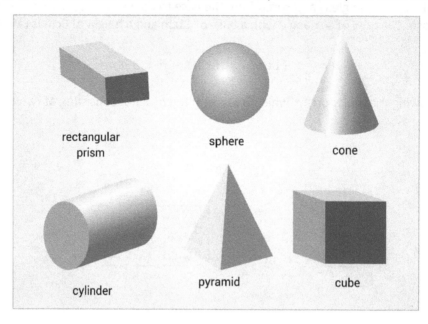

Cube
The cube is the simplest figure for which volume can be determined because all dimensions in a cube are equal. In the following figure, the length, width, and height of the cube are all represented by the variable *a* because these measurements are equal lengths.

The volume of any rectangular, three-dimensional object is found by multiplying its length by its width by its height. In the case of a cube, the length, width, and height are all equal lengths, represented by the variable *a*. Therefore, the equation used to calculate the volume is $(a \times a \times a)$ or a^3. In a real-world example of this situation, if the length of a side of the cube is 3 centimeters, the volume is calculated by utilizing the formula $(3 \times 3 \times 3) = 9$ cm³.

Rectangular Prism
The dimensions of a rectangular prism are not necessarily equal as those of a cube. Therefore, the formula for a rectangular prism recognizes that the dimensions vary and use different variables to represent these lengths. The length, width, and height of a rectangular prism are represented with the variables *a, b*, and *c*.

The equation used to calculate volume is length times width times height. Using the variables in the diagram above, this means $a \times b \times c$. In a real-world application of this situation, if *a*=2 cm, *b*=3 cm, and *c*=4 cm, the volume is calculated by utilizing the formula $3 \times 4 \times 5 = 60$ cm³.

Cylinder
Discovering a cylinder's volume requires the measurement of the cylinder's base, length of the radius, and height. The height of the cylinder can be represented with variable *h*, and the radius can be represented with variable *r*.

The formula to find the volume of a cylinder is $\pi r^2 h$. Notice that πr^2 is the formula for the area of a circle. This is because the base of the cylinder is a circle. To calculate the volume of a cylinder, the slices of circles needed to build the entire height of the cylinder are added together. For example, if the radius is 5 feet and the height of the cylinder is 10 feet, the cylinder's volume is calculated by using the following equation: $\pi 5^2 \times 10$. Substituting 3.14 for π, the volume is 785.4 ft³.

Pyramid
To calculate the volume of a pyramid, the area of the base of the pyramid is multiplied by the pyramid's height by $\frac{1}{3}$. The area of the base of the pyramid is found by multiplying the base length by the base width.

Therefore, the formula to calculate a pyramid's volume is $(L \times W \times H) \div 3$.

Cone
The formula to calculate the volume of a circular cone is similar to the formula for the volume of a pyramid. The primary difference in determining the area of a cone is that a circle serves as the base of a cone. Therefore, the area of a circle is used for the cone's base.

The variable *r* represents the radius, and the variable *h* represents the height of the cone. The formula used to calculate the volume of a cone is $\frac{1}{3}\pi r^2 h$. Essentially, the area of the base of the cone is multiplied by the cone's height. In a real-life example where the radius of a cone is 2 meters and the height of a cone is 5 meters, the volume of the cone is calculated by utilizing the formula:

$$\frac{1}{3}\pi 2^2 \times 5 = 21$$

After substituting 3.14 for π, the volume is 785.4 ft³.

Sphere
The volume of a sphere uses π due to its circular shape.

The length of the radius, *r*, is the only variable needed to determine the sphere's volume. The formula to calculate the volume of a sphere is $\frac{4}{3}\pi r^3$. Therefore, if the radius of a sphere is 8 centimeters, the volume of the sphere is calculated by utilizing the formula:

$$\frac{4}{3}\pi(8)^3 = 2{,}143 \; cm^3$$

Symmetry

Symmetry is another concept in geometry. If a two-dimensional shape can be folded along a straight line and the halves line up exactly, the figure is **symmetric.** The line is known as a **line of symmetry**. Circles, squares, and rectangles are examples of symmetric shapes.

Below is an example of a pentagon with a line of symmetry drawn.

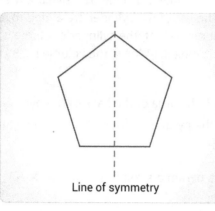

Line of symmetry

If a line cannot be drawn anywhere on the object to flip the figure onto itself, the object is **asymmetrical**. An example of a shape with no line of symmetry would be a scalene triangle.

If an object can be rotated about its center to any degree smaller than 360, and it lies directly on top of itself, the object is said to have **rotational symmetry**. An example of this type of symmetry is shown below. The pentagon has an order of 5.

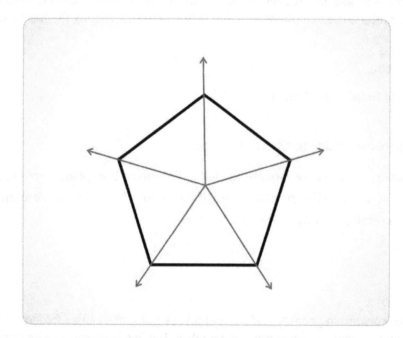

The rotational symmetry lines in the figure above can be used to find the angles formed at the center of the pentagon. Knowing that all of the angles together form a full circle, at 360 degrees, the figure can be split into 5 angles equally. By dividing the 360° by 5, each angle is 72°.

Given the length of one side of the figure, the perimeter of the pentagon can also be found using rotational symmetry. If one side length was 3 cm, that side length can be rotated onto each other side length four times. This would give a total of 5 side lengths equal to 3 cm. To find the perimeter, or distance around the figure, multiply 3 by 5. The perimeter of the figure would be 15 cm.

If a line cannot be drawn anywhere on the object to flip the figure onto itself or rotated less than or equal to 180 degrees to lay on top of itself, the object is asymmetrical. Examples of these types of figures are shown below.

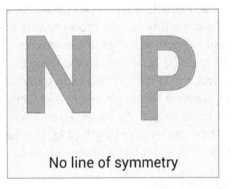

Measuring Lengths of Objects

The length of an object can be measured using standard tools such as rulers, yard sticks, meter sticks, and measuring tapes. The following image depicts a yardstick:

Choosing the right tool to perform the measurement requires determining whether United States customary units or metric units are desired, and having a grasp of the approximate length of each unit and the approximate length of each tool. The measurement can still be performed by trial and error without the knowledge of the approximate size of the tool.

For example, to determine the length of a room in feet, a United States customary unit, various tools can be used for this task. These include a ruler (typically 12 inches/1 foot long), a yardstick (3 feet/1 yard long), or a tape measure displaying feet (typically either 25 feet or 50 feet). Because the length of a room is much larger than the length of a ruler or a yardstick, a tape measure should be used to perform the measurement.

When the correct measuring tool is selected, the measurement is performed by first placing the tool directly above or below the object (if making a horizontal measurement) or directly next to the object (if making a vertical measurement). The next step is aligning the tool so that one end of the object is at the mark for zero units, then recording the unit of the mark at the other end of the object. To give the length of a paperclip in metric units, a ruler displaying centimeters is aligned with one end of the paper clip to the mark for zero centimeters.

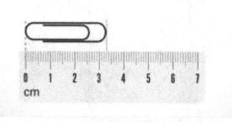

Directly down from the other end of the paperclip is the mark that measures its length. In this case, that mark is two small dashes past the 3 centimeter mark. Each small dash is 1 millimeter (or .1 centimeters). Therefore, the length of the paper clip is 3.2 centimeters.

To compare the lengths of objects, each length must be expressed in the same unit. If possible, the objects should be measured with the same tool or with tools utilizing the same units. For example, a ruler and a yardstick can both measure length in inches. If the lengths of the objects are expressed in different units, these different units must be converted to the same unit before comparing them. If two lengths are expressed in the same unit, the lengths may be compared by subtracting the smaller value from the larger value. For example, suppose the lengths of two gardens are to be compared. Garden A has a length of 4 feet, and garden B has a length of 2 yards. 2 yards is converted to 6 feet so that the measurements have similar units. Then, the smaller length (4 feet) is subtracted from the larger length (6ft): 6ft – 4ft = 2ft. Therefore, garden B is 2 feet larger than garden A.

Identifying Relative Sizes of Measurement Units

Measurement is how an object's length, width, height, weight, and so on, are quantified. Measurement is related to counting, but it is a more refined process.

The United States customary system and the metric system each consist of distinct units to measure lengths and volume of liquids. The U.S. customary units for length, from smallest to largest, are: inch (in), foot (ft), yard (yd), and mile (mi). The metric units for length, from smallest to largest, are: millimeter (mm), centimeter (cm), decimeter (dm), meter (m), and kilometer (km). The relative size of each unit of length is shown below.

U.S. Customary	Metric	Conversion
12in = 1ft	10mm = 1cm	1in = 2.54cm
36in = 3ft = 1yd	10cm = 1dm(decimeter)	1m ≈ 3.28ft ≈ 1.09yd
5,280ft = 1,760yd = 1mi	100cm = 10dm = 1m	1mi ≈ 1.6km
	1000m = 1km	

The U.S. customary units for volume of liquids, from smallest to largest, are: fluid ounces (fl oz), cup (c), pint (pt), quart (qt), and gallon (gal). The metric units for volume of liquids, from smallest to largest, are: milliliter (mL), centiliter (cL), deciliter (dL), liter (L), and kiloliter (kL). The relative size of each unit of liquid volume is shown below.

U.S. Customary	Metric	Conversion
8fl oz = 1c	10mL = 1cL	1pt ≈ 0.473L
2c = 1pt	10cL = 1dL	1L ≈ 1.057qt
4c = 2pt = 1qt	1,000mL = 100cL = 10dL = 1L	1gal ≈ 3.785L
4qt = 1gal	1,000L = 1kL	

The U.S. customary system measures weight (how strongly Earth is pulling on an object) in the following units, from least to greatest: ounce (oz), pound (lb), and ton. The metric system measures mass (the quantity of matter within an object) in the following units, from least to greatest: milligram (mg), centigram (cg), gram (g), kilogram (kg), and metric ton (MT).

The relative sizes of each unit of weight and mass are shown below.

U.S. Measures of Weight	Metric Measures of Mass
16oz = 1lb	10mg = 1cg
2,000lb = 1 ton	100cg = 1g
	1,000g = 1kg
	1,000kg = 1MT

Note that weight and mass DO NOT measure the same thing.

Time is measured in the following units, from shortest to longest: second (sec), minute (min), hour (h), day (d), week (wk), month (mo), year (yr), decade, century, millennium. The relative sizes of each unit of time is shown below.

- 60sec = 1min
- 60min = 1h
- 24hr = 1d
- 7d = 1wk
- 52wk = 1yr
- 12mo = 1yr
- 10yr = 1 decade
- 100yrs = 1 century
- 1,000yrs = 1 millennium

Converting Measurements

When working with different systems of measurement, conversion from one unit to another may be necessary. The conversion rate must be known to convert units. One method for converting units is to write and solve a proportion. The arrangement of values in a proportion is extremely important. Suppose that your doctor measures your height in inches to be 51 inches and you want to know if you will be tall enough to ride the big roller coasters, which require a height of 4 feet. Therefore, the problem requires converting your height in inches to feet.

Looking at the conversion chart, it can be seen that 1 foot = 12 inches. A proportion can be set up using this conversion and *x* representing the missing value.

The proportion can be written in any of the following ways:

$$\frac{1}{12} = \frac{x}{51} \left(\frac{feet\ for\ conversion}{inches\ for\ conversion} = \frac{unknown\ feet}{inches\ given} \right)$$

$$\frac{12}{1} = \frac{51}{x} \left(\frac{inches\ for\ conversion}{feet\ for\ conversion} = \frac{inches\ given}{unknown\ feet} \right)$$

$$\frac{1}{x} = \frac{12}{51} \left(\frac{feet\ for\ conversion}{unknown\ feet} = \frac{inches\ for\ conversion}{inches\ given} \right)$$

$$\frac{x}{1} = \frac{51}{12} \left(\frac{unknown\ feet}{feet\ for\ conversion} = \frac{inches\ given}{inches\ for\ conversion} \right)$$

To solve a proportion, the ratios are cross-multiplied and the resulting equation is solved. When cross-multiplying, all four proportions above will produce the same equation: $(12)(x) = (51)(1) \rightarrow 12x = 51$. Dividing by 12 to isolate the variable *x*, the result is *x* = 4.25. The variable *x* represented the unknown number of feet. Therefore, the conclusion is that 51 inches converts (is equal) to 4.25 feet, meaning you are tall enough to ride the roller coasters!

Note that while there are four options presented for how to correctly set up the proportion, not every arrangement is correct. It would be incorrect, for example to set up the following:

$$\frac{1}{12} = \frac{51}{x} \left(\frac{feet\ for\ conversion}{inches\ for\ conversion} = \frac{inches\ given}{unknown\ feet} \right)$$

This is because when cross-multiplying, it would be seen that inches would be multiplied by inches and feet by feet instead of inches with feet. Therefore, the units would not cancel out and the proportion is incorrect.

Let's try another problem. Suppose that a problem requires converting 20 fluid ounces to cups. To do so, a proportion can be written using the conversion rate of 8fl oz = 1c with *x* representing the missing value. Again, the proportion can be written in any of the following ways:

$$\frac{1}{8} = \frac{x}{20} \left(\frac{c\ for\ conversion}{fl\ oz\ for\ conversion} = \frac{unknown\ c}{fl\ oz\ given} \right)$$

$$\frac{8}{1} = \frac{20}{x} \left(\frac{fl\ oz\ for\ conversion}{c\ for\ conversion} = \frac{fl\ oz\ given}{unknown\ c} \right)$$

$$\frac{1}{x} = \frac{8}{20} \left(\frac{c\ for\ conversion}{unknown\ c} = \frac{fl\ oz\ for\ conversion}{fl\ oz\ given} \right)$$

$$\frac{x}{1} = \frac{20}{8} \left(\frac{unknown\ c}{c\ for\ conversion} = \frac{fl\ oz\ given}{fl\ oz\ for\ conversion} \right)$$

To solve the proportion, the ratios are cross-multiplied and the resulting equation is solved. When cross-multiplying, all four proportions above will produce the same equation: $(8)(x) = (20)(1) \rightarrow 8x = 20$. Dividing by 8 to isolate the variable x, the result is $x = 2.5$. The variable x represented the unknown number of cups. Therefore, the conclusion is that 20 fluid ounces converts (is equal) to 2.5 cups.

Sometimes converting units requires writing and solving more than one proportion. Suppose an exam question asks to determine how many hours are in 2 weeks. Without knowing the conversion rate between hours and weeks, this can be determined knowing the conversion rates between weeks and days, and between days and hours. First, weeks are converted to days, then days are converted to hours. To convert from weeks to days, the following proportion can be written:

$$\frac{7}{1} = \frac{x}{2} \left(\frac{days\ conversion}{weeks\ conversion} = \frac{days\ unknown}{weeks\ given} \right)$$

Cross-multiplying produces:

$$(7)(2) = (x)(1) \rightarrow 14 = x$$

Therefore, 2 weeks is equal to 14 days. Next, a proportion is written to convert 14 days to hours:

$$\frac{24}{1} = \frac{x}{14} \left(\frac{conversion\ hours}{conversion\ days} = \frac{unknown\ hours}{given\ days} \right)$$

Cross-multiplying produces:

$$(24)(14) = (x)(1)$$

$$336 = x$$

Therefore, the answer is that there are 336 hours in 2 weeks.

Solving Problems Concerning Measurements

Problems that involve measurements of length, time, volume, etc. are generally dependent upon understanding how to manipulate between various units of measurement, as well as understanding their equivalencies.

Determining Solutions to Problems Involving Time
Time is measured in units such as *seconds, minutes, hours, days,* and *years*. For example, there are 60 seconds in a minute, 60 minutes in each hour, and 24 hours in a day.

When dealing with problems involving elapsed time, break the problem down into workable parts. For example, suppose the length of time between 1:15pm and 3:45pm must be determined. From 1:15pm to 2:00pm is 45 minutes (knowing there are 60 minutes in an hour). From 2:00pm to 3:00pm is 1 hour. From 3:00pm to 3:45pm is 45 minutes. The total elapsed time is 45 minutes plus 1 hour plus 45 minutes. This sum produces 1 hour and 90 minutes. 90 minutes is over an hour, so this is converted to 1 hour (60 minutes) and 30 minutes. The total elapsed time can now be expressed as 2 hours and 30 minutes.

To illustrate time intervals, a clock face can show solutions.

For example, Ani needs to complete all of her chores by 1:50 p.m. If she begins her chores at 1:00 p.m., can she finish the following? Vacuuming (15 minutes), dusting (10 minutes), replacing light bulbs (5 minutes), and degreasing the garage floor (25 minutes).

A blank clock face is useful in illustrating the time lapse necessary for all of Ani's tasks.

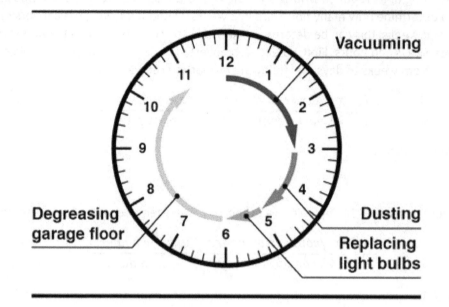

It is easy to see that the chores will span beyond 50 minutes after the hour, so no, Ani could not complete the chores in the given time frame.

When dealing with problems involving elapsed time, breaking the problem down into workable parts is helpful. For example, suppose the length of time between 1:15pm and 3:45pm must be determined. From 1:15pm to 2:00pm is 45 minutes (knowing there are 60 minutes in an hour). From 2:00pm to 3:00pm is 1 hour. From 3:00pm to 3:45pm is 45 minutes. The total elapsed time is 45 minutes plus 1 hour plus 45 minutes. This sum produces 1 hour and 90 minutes. 90 minutes is over an hour, so this is converted to 1 hour (60 minutes) and 30 minutes. The total elapsed time can now be expressed as 2 hours and 30 minutes.

Determining Solutions to Problems Involving Money
Let's consider a problem involving change. Gwen wants to purchase a number of items at the school supply store. Prices are as follows:

Erasers = $0.05
Highlighters = $0.20
Pencils = $0.10
Pens = $0.15

Gwen wants to buy 1 highlighter, 5 erasers, 2 pencils, and 3 pens. Would she be able to use the following coins, and if so, what are possible combinations Gwen could use?

To begin, add the total value of the coins:

2 quarters + 5 dimes + 2 nickel

$$(2 \times \$0.25) + (5 \times \$0.10) + (2 \times \$0.05) = \$1.10$$

Then calculate the total cost of the items Gwen wants to purchase:

Erasers = $0.05
Highlighters = $0.20
Pencils = $0.10
Pens = $0.15

Gwen wants to buy 1 highlighter, 5 erasers, 2 pencils, and 3 pens.

1 highlighter + 5 erasers + 2 pencils + 2 pens

$$\$0.20 + (5 \times \$0.05) + (2 \times \$0.10) + (2 \times \$0.15)$$

$$\$0.20 + \$0.25 + \$0.20 + \$0.30 = \$0.95$$

Gwen has enough money to purchase all of the items, and there is only one combination that would provide the correct amount of $0.95: 2 quarters, 4 dimes, and 1 nickel.

Here is another example of a problem involving money that is a bit more difficult:

A store is having a spring sale, where everything is 70% off. You have $45.00 to spend. A jacket is regularly priced at $80.00. Do you have enough to buy the jacket and a pair of gloves, regularly priced at $20.00?

There are two ways to approach this.

Method 1:

Set up the equations to find the sale prices: the original price minus the amount discounted.
$80.00 - ($80.00 (0.70)) = sale cost of the jacket.
$20.00 – ($20.00 (0.70)) = sale cost of the gloves.
Solve for the sale cost.
$24.00 = sale cost of the jacket.
$6.00 = sale cost of the gloves.
Determine if you have enough money for both.
$24.00 + $6.00 = total sale cost.
$30.00 is less than $45.00, so you can afford to purchase both.

Method 2:

Determine the percent of the original price that you will pay.

100% − 70% = 30%

Set up the equations to find the sale prices.

$80.00 (0.30) = cost of the jacket.

$20.00 (0.30) = cost of the gloves.

Solve.

$24.00 = cost of the jacket.

$6.00 = cost of the gloves.

Determine if you have enough money for both.

$24.00 + $6.00 = total sale cost.

$30.00 is less than $45.00, so you can afford to purchase both.

<u>Determining Solutions to Problems Involving Length</u>

Identifying and utilizing the proper units for the scenario requires knowing how to apply the conversion rates for length. For example, given a scenario that requires subtracting 8 inches from $2\frac{1}{2}$ feet, both values should first be expressed in the same unit (they could be expressed $\frac{2}{3}$ft & $2\frac{1}{2}$ft, or 8in and 30in). The desired unit for the answer may also require converting back to another unit.

Consider the following scenario that involves length: A parking area along the river is only wide enough to fit one row of cars and is $\frac{1}{2}$ kilometers long. The average space needed per car is 5 meters. How many cars can be parked along the river? First, all measurements should be converted to similar units: $\frac{1}{2}$km = 500m. The operation(s) needed should be identified. Because the problem asks for the number of cars, the total space should be divided by the space per car. 500 meters divided by 5 meters per car yields a total of 100 cars. Written as an expression, the meters unit cancels and the cars unit is left: $\frac{500m}{5m/car}$ the same as $500m \times \frac{1\ car}{5m}$ yields 100 cars.

<u>Determining Solutions to Problems Involving Volume</u>

Volume is how much space something occupies. That "something" can be liquid or solid. Essentially, volume is a measurement of capacity. Whereas area is calculated by counting squares within a two-dimensional object, volume is calculated by counting cubes within a three-dimensional object. It is a measure of the space a figure occupies. Volume is measured using cubic units, such as cubic inches, feet, centimeters, or kilometers. Centimeter cubes can be utilized in the classrooms in order to promote understanding of volume.

For instance, if 10 cubes were placed along the length of a rectangle, with 8 cubes placed along its width, and the remaining area was filled in with cubes, there would 80 cubes in total, which would equal a volume of 80 cubic centimeters. Its area would equal 80 square centimeters. If that shape was doubled so that its height consists of two cube lengths, there would be 160 cubes, and its volume would be 160 cubic centimeters. Adding another level of cubes would mean that there would be $3 \times 80 = 240$ cubes. This idea shows that volume is calculated by multiplying area times height. The actual formula for volume of a three-dimensional rectangular solid is $V = l \times w \times h$, where *l* represents length, *w* represents width, and *h* represents height. Volume can also be thought of as area of the base times the height. The base in this case would be the entire rectangle formed by *l* and *w*. Here is an example of a rectangular solid with labeled sides:

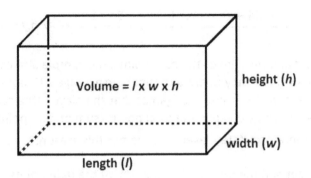

Volume = *l* x *w* x *h*

height (*h*)

width (*w*)

length (*l*)

A **cube** is a special type of rectangular solid in which its length, width, and height are the same. If this length is *s*, then the formula for the volume of a cube is $V = s \times s \times s$.

The U.S. customary units for volume of liquids, from smallest to largest, are: fluid ounces (fl oz), cup (c), pint (pt), quart (qt), and gallon (gal). The metric units for volume of liquids, from smallest to largest, are: milliliter (mL), centiliter (cL), deciliter (dL), liter (L), and kiloliter (kL.).

A problem involving liquid volume would be: If Mart needed 2 quarts of liquid for a recipe and only has a measuring cup, how could he measure out 2 quarts?

The solution would involve Mart measuring out 2 quarts by filling the cup 8 times.

Determining Solutions to Problems Involving Mass
The metric system measures **mass**, which is the quantity of matter within an object. Mass and weight do not measure the same thing. **Weight** is affected by gravity, and deals with how strongly Earth is pulling on an object.

The following is an example of a problem involving mass:

A piggy bank contains 12 dollars' worth of nickels. The mass of a nickel is 5 grams, and the empty piggy bank has a mass of 1050 grams. What is the total mass of the full piggy bank?

A dollar contains 20 nickels. Therefore, if there are 12 dollars' worth of nickels, there are $12 \times 20 = 240$ nickels. The mass of each nickel is 5 grams. Therefore, the mass of the nickels is $240 \times 5 = 1,200$ grams. Adding in the mass of the empty piggy bank, the mass of the filled bank 2,250 grams.

Data Analysis & Probability

Measures of Center and Range

The center of a set of data (statistical values) can be represented by its mean, median, or mode. These are sometimes referred to as measures of central tendency. The **mean** is the average of the data set. The mean can be calculated by adding the data values and dividing by the sample size (the number of data points). Suppose a student has test scores of 93, 84, 88, 72, 91, and 77. To find the mean, or average, the scores are added and the sum is divided by 6 because there are 6 test scores:

$$\frac{93 + 84 + 88 + 72 + 91 + 77}{6} = \frac{505}{6} = 84.17$$

Given the mean of a data set and the sum of the data points, the **sample size** can be determined by dividing the sum by the mean. Suppose you are told that Kate averaged 12 points per game and scored a total of 156 points for the season. The number of games that she played (the sample size or the number of data points) can be determined by dividing the total points (sum of data points) by her average (mean of data points): $\frac{156}{12} = 13$. Therefore, Kate played in 13 games this season.

If given the mean of a data set and the sample size, the sum of the data points can be determined by multiplying the mean and sample size. Suppose you are told that Tom worked 6 days last week for an average of 5.5 hours per day. The total number of hours worked for the week (sum of data points) can be determined by multiplying his daily average (mean of data points) by the number of days worked (sample size): $5.5 \times 6 = 33$. Therefore, Tom worked a total of 33 hours last week.

The **median** of a data set is the value of the data point in the middle when the sample is arranged in numerical order. To find the median of a data set, the values are written in order from least to greatest. The lowest and highest values are simultaneously eliminated, repeating until the value in the middle remains. Suppose the salaries of math teachers are: $35,000; $38,500; $41,000; $42,000; $42,000; $44,500; $49,000. The values are listed from least to greatest to find the median. The lowest and highest values are eliminated until only the middle value remains. Repeating this step three times reveals a median salary of $42,000. If the sample set has an even number of data points, two values will remain after all others are eliminated. In this case, the mean of the two middle values is the median. Consider the following data set: 7, 9, 10, 13, 14, 14. Eliminating the lowest and highest values twice leaves two values, 10 and 13, in the middle. The mean of these values $\left(\frac{10+13}{2}\right)$ is the median. Therefore, the set has a median of 11.5.

The **mode** of a data set is the value that appears most often. A data set may have a single mode, multiple modes, or no mode. If different values repeat equally as often, multiple modes exist. If no value repeats, no mode exists. Consider the following data sets:

- A: 7, 9, 10, 13, 14, 14
- B: 37, 44, 33, 37, 49, 44, 51, 34, 37, 33, 44
- C: 173, 154, 151, 168, 155

Set A has a mode of 14. Set B has modes of 37 and 44. Set C has no mode.

The **range** of a data set is the difference between the highest and the lowest values in the set. The range can be considered the span of the data set. To determine the range, the smallest value in the set is subtracted from the largest value. The ranges for the data sets A, B, and C above are calculated as follows:

A: $14 - 7 = 7$

B: $51 - 33 = 18$

C: $173 - 151 = 22$

Best Description of a Set of Data

Measures of central tendency, namely mean, median, and mode, describe characteristics of a set of data. Specifically, they are intended to represent a *typical* value in the set by identifying a central position of the set. Depending on the characteristics of a specific set of data, different measures of central tendency are more indicative of a typical value in the set.

When a data set is grouped closely together with a relatively small range and the data is spread out somewhat evenly, the mean is an effective indicator of a typical value in the set. Consider the following data set representing the height of sixth grade boys in inches: 61 inches, 54 inches, 58 inches, 63 inches, 58 inches. The mean of the set is 58.8 inches. The data set is grouped closely (the range is only 9 inches) and the values are spread relatively evenly (three values below the mean and two values above the mean). Therefore, the mean value of 58.8 inches is an effective measure of central tendency in this case.

When a data set contains a small number of values either extremely large or extremely small when compared to the other values, the mean is not an effective measure of central tendency. Consider the following data set representing annual incomes of homeowners on a given street: $71,000; $74,000; $75,000; $77,000; $340,000. The mean of this set is $127,400. This figure does not indicate a typical value in the set, which contains four out of five values between $71,000 and $77,000. The median is a much more effective measure of central tendency for data sets such as these. Finding the middle value diminishes the influence of outliers, or numbers that may appear out of place, like the $340,000 annual income. The median for this set is $75,000 which is much more typical of a value in the set.

The mode of a data set is a useful measure of central tendency for categorical data when each piece of data is an option from a category. Consider a survey of 31 commuters asking how they get to work with results summarized below.

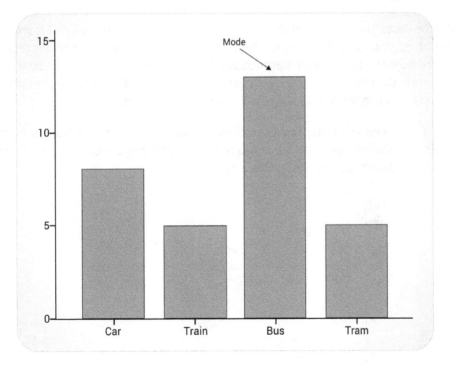

The mode for this set represents the value, or option, of the data that repeats most often. This indicates that the bus is the most popular method of transportation for the commuters.

Effects of Changes in Data

Changing all values of a data set in a consistent way produces predictable changes in the measures of the center and range of the set. A linear transformation changes the original value into the new value by either adding a given number to each value, multiplying each value by a given number, or both. Adding (or subtracting) a given value to each data point will increase (or decrease) the mean, median, and any modes by the same value. However, the range will remain the same due to the way that range is calculated. Multiplying (or dividing) a given value by each data point will increase (or decrease) the mean, median, and any modes, and the range by the same factor.

Consider the following data set, call it set P, representing the price of different cases of soda at a grocery store: $4.25, $4.40, $4.75, $4.95, $4.95, $5.15. The mean of set P is $4.74. The median is $4.85. The mode of the set is $4.95. The range is $0.90. Suppose the state passes a new tax of $0.25 on every case of soda sold. The new data set, set T, is calculated by adding $0.25 to each data point from set P. Therefore, set T consists of the following values: $4.50, $4.65, $5.00, $5.20, $5.20, $5.40. The mean of set T is $4.99. The median is $5.10. The mode of the set is $5.20. The range is $.90. The mean, median and mode of set T is equal to $0.25 added to the mean, median, and mode of set P. The range stays the same.

Now suppose, due to inflation, the store raises the cost of every item by 10 percent. Raising costs by 10 percent is calculated by multiplying each value by 1.1. The new data set, set I, is calculated by multiplying each data point from set T by 1.1. Therefore, set I consists of the following values: $4.95, $5.12, $5.50, $5.72, $5.72, $5.94. The mean of set I is $5.49. The median is $5.61. The mode of the set is $5.72. The range is $0.99. The mean, median, mode, and range of set I is equal to 1.1 multiplied by the mean, median, mode, and range of set T because each increased by a factor of 10 percent.

Describing a Set of Data

A set of data can be described in terms of its center, spread, shape and any unusual features. The center of a data set can be measured by its mean, median, or mode. The **spread** of a data set refers to how far the data points are from the center (mean or median). The spread can be measured by the range or the quartiles and interquartile range. A data set with data points clustered around the center will have a small spread. A data set covering a wide range will have a large spread.

When a data set is displayed as a histogram or frequency distribution plot, the shape indicates if a sample is normally distributed, symmetrical, or has measures of skewness or kurtosis. When graphed, a data set with a normal distribution will resemble a bell curve.

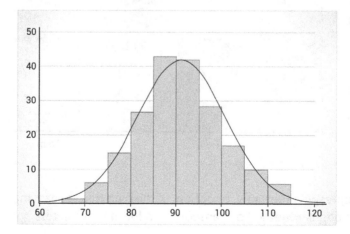

If the data set is symmetrical, each half of the graph when divided at the center is a mirror image of the other. If the graph has fewer data points to the right, the data is skewed right. If it has fewer data points to the left, the data is skewed left.

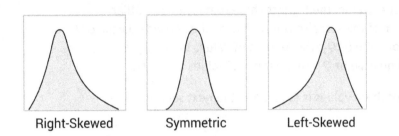

Right-Skewed   Symmetric   Left-Skewed

Kurtosis is a measure of whether the data is heavy-tailed with a high number of outliers, or light-tailed with a low number of outliers.

A description of a data set should include any unusual features such as gaps or outliers. A **gap** is a span within the range of the data set containing no data points. An **outlier** is a data point with a value either extremely large or extremely small when compared to the other values in the set.

Interpreting Displays of Data

A set of data can be visually displayed in various forms allowing for quick identification of characteristics of the set. **Histograms,** such as the one shown below, display the number of data points (or the frequency) on the vertical axis that fall into given intervals (horizontal axis) across the range of the set. Suppose the histogram below displays the number of black cherry trees of different heights in a certain orchard. Histograms can describe the center, spread, shape, and any unusual characteristics of a data set.

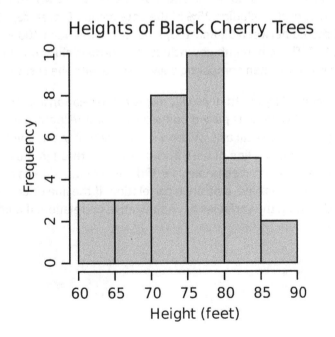

A **box plot**, also called a **box-and-whisker plot**, divides the data points into four groups and displays the five-number summary for the set, as well as any outliers. The five-number summary consists of:

- The lower extreme: the lowest value that is not an outlier
- The higher extreme: the highest value that is not an outlier
- The median of the set: also referred to as the second quartile or Q_2
- The first quartile or Q_1: the median of values below Q_2
- The third quartile or Q_3: the median of values above Q_2

Calculating each of these values is covered in the next section.

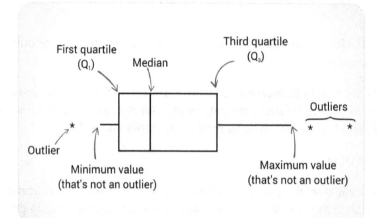

Suppose the box plot displays IQ scores for 12th grade students at a given school. The five number summary of the data consists of: lower extreme (67); upper extreme (127); Q_2 or median (100); Q_1 (91); Q_3 (108); and outliers (135 and 140). Although all data points are not known from the plot, the points are divided into four quartiles each, including 25% of the data points. Therefore, 25% of students scored between 67 and 91, 25% scored between 91 and 100, 25% scored between 100 and 108, and 25% scored between 108 and 127. These percentages include the normal values for the set and exclude the outliers. This information is useful when comparing a given score with the rest of the scores in the set.

A **scatter plot** is a mathematical diagram that visually displays the relationship or connection between two variables. The independent variable is placed on the x-axis, or horizontal axis, and the dependent variable is placed on the y-axis, or vertical axis. When visually examining the points on the graph, if the points model a linear relationship, or a line of best-fit can be drawn through the points with the points relatively close on either side, then a correlation exists. If the line of best-fit has a positive slope (rises from left to right), then the variables have a positive correlation. If the line of best-fit has a negative slope (falls from left to right), then the variables have a negative correlation. If a line of best-fit cannot be drawn, then no correlation exists. A positive or negative correlation can be categorized as strong or weak, depending on how closely the points are graphed around the line of best-fit.

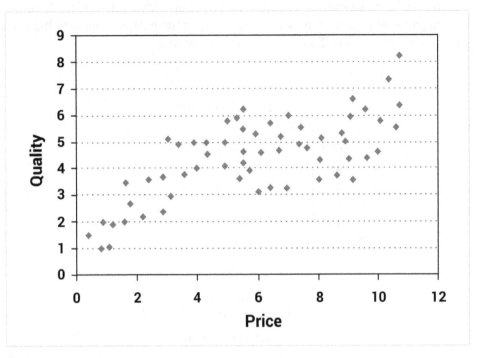

Graphical Representation of Data

Various graphs can be used to visually represent a given set of data. Each type of graph requires a different method of arranging data points and different calculations of the data. Examples of histograms, box plots, and scatter plots are discussed in the previous section, *Interpreting Displays of Data*. To construct a histogram, the range of the data points is divided into equal intervals. The frequency for each interval is then determined, which reveals how many points fall into each interval. A graph is constructed with the vertical axis representing the frequency and the horizontal axis representing the intervals. The lower value of each interval should be labeled along the horizontal axis. Finally, for each interval, a bar is drawn from the lower value of each interval to the lower value of the next interval with a height equal to the frequency of the interval. Because of the intervals, histograms do not have any gaps between bars along the horizontal axis.

As mentioned, a **scatter plot** displays the relationship between two variables. Values for the independent variable, typically denoted by *x*, are paired with values for the dependent variable, typically denoted by *y*. Each set of corresponding values are written as an ordered pair (*x*, *y*). To construct the graph, a coordinate grid is labeled with the *x*-axis representing the independent variable and the *y*-axis representing the dependent variable. Each ordered pair is graphed.

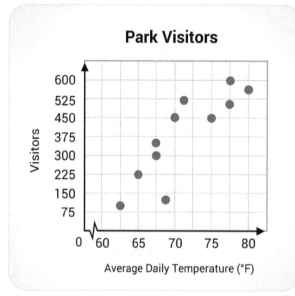

Like a scatter plot, a **line graph** compares variables that change continuously, typically over time. Paired data values (ordered pair) are plotted on a coordinate grid with the *x*- and *y*-axis representing the variables. A line is drawn from each point to the next, going from left to right. The line graph below displays cell phone use for given years (two variables) for men, women, and both sexes (three data sets).

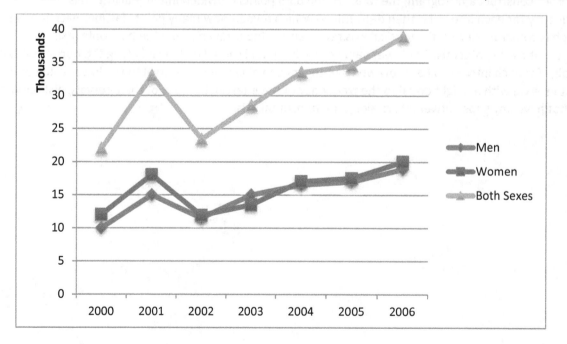

A **line plot**, also called **dot plot**, displays the frequency of data (numerical values) on a number line. To construct a line plot, a number line is used that includes all unique data values. It is marked with x's or dots above the value the number of times that the value occurs in the data set.

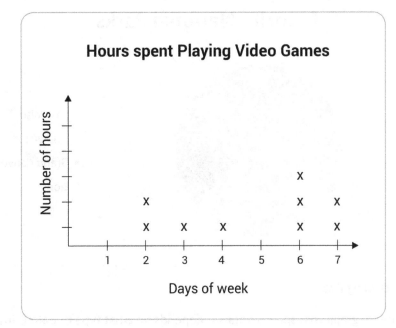

A **bar graph** looks similar to a histogram but displays categorical data. The horizontal axis represents each category and the vertical axis represents the frequency for the category. A bar is drawn for each category (often different colors) with a height extending to the frequency for that category within the data set. A double bar graph displays two sets of data that contain data points consisting of the same categories. The double bar graph below indicates that two girls and four boys like Pad Thai the most out of all the foods, two boys and five girls like pizza, and so on.

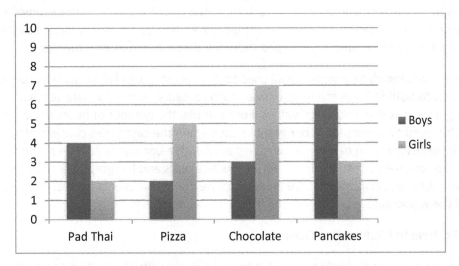

A **circle graph**, also called a **pie chart**, displays categorical data with each category representing a percentage of the whole data set. To construct a circle graph, the percent of the data set for each category must be determined. To do so, the frequency of the category is divided by the total number of data points and converted to a percent. For example, if 80 people were asked their favorite pizza topping and 20 responded cheese, then cheese constitutes 25% of the data ($\frac{20}{80} = .25 = 25\%$). Each category in a data set is represented by a *slice* of the circle proportionate to its percentage of the whole.

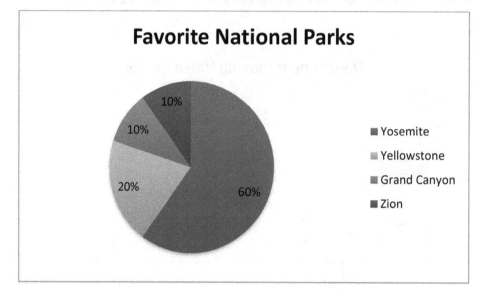

Choice of Graphs to Display Data

Choosing the appropriate graph to display a data set depends on what type of data is included in the set and what information must be displayed. Histograms and box plots can be used for data sets consisting of individual values across a wide range. Examples include test scores and incomes. Histograms and box plots will indicate the center, spread, range, and outliers of a data set. A histogram will show the shape of the data set, while a box plot will divide the set into quartiles (25% increments), allowing for comparison between a given value and the entire set.

Scatter plots and line graphs can be used to display data consisting of two variables. Examples include height and weight, or distance and time. A correlation between the variables is determined by examining the points on the graph. Line graphs are used if each value for one variable pairs with a distinct value for the other variable. Line graphs show relationships between variables.

Line plots, bar graphs, and circle graphs are all used to display categorical data, such as surveys. Line plots and bar graphs both indicate the frequency of each category within the data set. A line plot is used when the categories consist of numerical values. For example, the number of hours of TV watched by individuals is displayed on a line plot. A bar graph is used when the categories consists of words. For example, the favorite ice cream of individuals is displayed with a bar graph. A circle graph can be used to display either type of categorical data. However, unlike line plots and bar graphs, a circle graph does not indicate the frequency of each category. Instead, the circle graph represents each category as its percentage of the whole data set.

Probabilities Relative to Likelihood of Occurrence

Probability is a measure of how likely an event is to occur. Probability is written as a fraction between zero and one. If an event has a probability of zero, the event will never occur. If an event has a probability of one, the event will definitely occur. If the probability of an event is closer to zero, the event is unlikely to occur. If the probability of an event is closer to one, the event is more likely to occur. For example, a probability of $\frac{1}{2}$ means that the event is equally as likely to occur as it is not to occur. An example of this is tossing a coin. To calculate the probability of an event, the number of favorable outcomes is divided by the number of total outcomes. For example, suppose you have 2 raffle tickets out of 20 total tickets sold. The probability that you win the raffle is calculated:

$$\frac{number\ of\ favorable\ outcomes}{total\ number of\ outcomes} = \frac{2}{20} = \frac{1}{10}\text{ (always reduce fractions)}$$

Therefore, the probability of winning the raffle is $\frac{1}{10}$ or 0.1.

Chance is the measure of how likely an event is to occur, written as a percent. If an event will never occur, the event has a 0% chance. If an event will certainly occur, the event has a 100% chance. If an event will sometimes occur, the event has a chance somewhere between 0% and 100%. To calculate chance, probability is calculated and the fraction is converted to a percent.

The probability of multiple events occurring can be determined by multiplying the probability of each event. For example, suppose you flip a coin with heads and tails, and roll a six-sided dice numbered one through six. To find the probability that you will flip heads AND roll a two, the probability of each event is determined and those fractions are multiplied. The probability of flipping heads is $\frac{1}{2}\left(\frac{1\ side\ with\ heads}{2\ sides\ total}\right)$ and the probability of rolling a two is:

$$\frac{1}{6}\left(\frac{1\ side\ with\ a\ 2}{6\ total\ sides}\right)$$

The probability of flipping heads AND rolling a 2 is:

$$\frac{1}{2} \times \frac{1}{6} = \frac{1}{12}$$

The above scenario with flipping a coin and rolling a dice is an example of independent events. Independent events are circumstances in which the outcome of one event does not affect the outcome of the other event. Conversely, dependent events are ones in which the outcome of one event affects the outcome of the second event. Consider the following scenario: a bag contains 5 black marbles and 5 white marbles. What is the probability of picking 2 black marbles without replacing the marble after the first pick?

The probability of picking a black marble on the first pick is:

$$\frac{5}{10}\left(\frac{5\ black\ marbles}{10\ total\ marbles}\right)$$

Assuming that a black marble was picked, there are now 4 black marbles and 5 white marbles for the second pick. Therefore, the probability of picking a black marble on the second pick is:

$$\frac{4}{9}\left(\frac{4\ black\ marbles}{9\ total\ marbles}\right)$$

To find the probability of picking two black marbles, the probability of each is multiplied:

$$\frac{5}{10} \times \frac{4}{9} = \frac{20}{90} = \frac{2}{9}$$

Practice Questions

1. Which of the following is equivalent to the value of the 3 in the number 792.134?
    a. 3×10

    b. 3×100

    c. $\frac{3}{10}$

    d. $\frac{3}{100}$

2. How will the following number be written in standard form:
$$(1 \times 10^4) + (3 \times 10^3) + (7 \times 10^1) + (8 \times 10^0)$$
    a. 137
    b. 13,078
    c. 1,378
    d. 8,731

3. How will the number 847.89632 be written if rounded to the nearest hundredth?
    a. 847.90
    b. 900
    c. 847.89
    d. 847.896

4. What is the value of the sum of $\frac{1}{3}$ and $\frac{2}{5}$?
    a. $\frac{3}{8}$

    b. $\frac{11}{15}$

    c. $\frac{11}{30}$

    d. $\frac{4}{5}$

5. What is the value of the expression: $7^2 - 3 \times (4 + 2) + 15 \div 5$?
    a. 10.4
    b. 28
    c. 34
    d. 58.2

6. How will $\frac{4}{5}$ be written as a percent?
    a. 40%
    b. 125%
    c. 90%
    d. 80%

7. If Danny takes 48 minutes to walk 3 miles, how long should it take him to walk 5 miles maintaining the same speed?
   a. 32 min
   b. 64 min
   c. 80 min
   d. 96 min

8. What are all the factors of 12?
   a. 12, 24, 36
   b. 1, 2, 4, 6, 12
   c. 12, 24, 36, 48
   d. 1, 2, 3, 4, 6, 12

9. A construction company is building a new housing development with the property of each house measuring 30 feet wide. If the length of the street is zoned off at 345 feet, how many houses can be built on the street?
   a. 11
   b. 115
   c. 11.5
   d. 12

10. How will the following algebraic expression be simplified: $(5x^2 - 3x + 4) - (2x^2 - 7)$?
    a. x^5
    b. $3x^2 - 3x + 11$
    c. $3x^2 - 3x - 3$
    d. $x - 3$

11. Kassidy drove for 3 hours at a speed of 60 miles per hour. Using the distance formula, $d = r \times t$ ($distance = rate \times time$), how far did Kassidy travel?
    a. 20 miles
    b. 180 miles
    c. 65 miles
    d. 120 miles

12. If $-3(x + 4) \geq x + 8$, what is the value of x?
    a. $x = 4$
    b. $x \geq 2$
    c. $x \geq -5$
    d. $x \leq -5$

13. Karen gets paid a weekly salary and a commission for every sale that she makes. The table below shows the number of sales and her pay for different weeks.

Sales	2	7		4	8
Pay	$380	$580		$460	$620

Which of the following equations represents Karen's weekly pay?
- a. $y = 90x + 200$
- b. $y = 90x - 200$
- c. $y = 40x + 300$
- d. $y = 40x - 300$

14. Which inequality represents the values displayed on the number line?

-5 -4 -3 -2 -1 0 1 2 3 4 5

- a. $x < 1$
- b. $x \leq 1$
- c. $x > 1$
- d. $x \geq 1$

15. What is the 42nd item in the pattern: ▲○○□ ▲○○□ ▲ ...?
- a. ○
- b. ▲
- c. □
- d. None of the above

16. Which of the following statements is true about the two lines below?

- a. The two lines are parallel but not perpendicular.
- b. The two lines are perpendicular but not parallel.
- c. The two lines are both parallel and perpendicular.
- d. The two lines are neither parallel nor perpendicular.

17. Which of the following figures is not a polygon?
- a. Decagon
- b. Cone
- c. Triangle
- d. Rhombus

18. What is the area of the regular hexagon shown below?

10.39

12

    a. 72
    b. 124.68
    c. 374.04
    d. 748.08

19. The area of a given rectangle is 24 centimeters. If the measure of each side is multiplied by 3, what is the area of the new figure?
    a. 48 cm
    b. 72 cm
    c. 216 cm
    d. 13,824 cm

20. What are the coordinates of the point plotted on the grid?

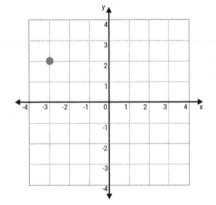

    a. (-3, 2)
    b. (2, -3)
    c. (-3, -2)
    d. (2, 3)

21. The perimeter of a 6-sided polygon is 56 cm. The length of three sides are 9 cm each. The length of two other sides are 8 cm each. What is the length of the missing side?
    a. 11 cm
    b. 12 cm
    c. 13 cm
    d. 10 cm

22. Katie works at a clothing company and sold 192 shirts over the weekend. $1/3$ of the shirts that were sold were patterned, and the rest were solid. Which mathematical expression would calculate the number of solid shirts Katie sold over the weekend?

    a. $192 \times \frac{1}{3}$

    b. $192 \div \frac{1}{3}$

    c. $192 \times (1 - \frac{1}{3})$

    d. $192 \div 3$

23. Which measure for the center of a small sample set is most affected by outliers?
    a. Mean
    b. Median
    c. Mode
    d. They are all equally affected

24. Given the value of a given stock at monthly intervals, which graph should be used to best represent the trend of the stock?
    a. Box plot
    b. Line plot
    c. Line graph
    d. Circle graph

25. What is the probability of randomly picking the winner and runner-up from a race of 4 horses and distinguishing which is the winner?

    a. $\frac{1}{4}$

    b. $\frac{1}{2}$

    c. $\frac{1}{16}$

    d. $\frac{1}{12}$

26. Which of the following numbers has the greatest value?
    a. 1.4378
    b. 1.07548
    c. 1.43592
    d. 0.89409

27. The value of 6 x 12 is the same as:
    a. 2 x 4 x 4 x 2
    b. 7 x 4 x 3
    c. 6 x 6 x 3
    d. 3 x 3 x 4 x 2

28. This chart indicates how many sales of CDs, vinyl records, and MP3 downloads occurred over the last year. Approximately what percentage of the total sales was from CDs?

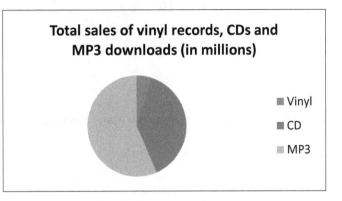

Total sales of vinyl records, CDs and MP3 downloads (in millions)

- Vinyl
- CD
- MP3

  a. 55%
  b. 25%
  c. 40%
  d. 5%

29. After a 20% sale discount, Frank purchased a new refrigerator for $850. How much did he save from the original price?
  a. $170
  b. $212.50
  c. $105.75
  d. $200

30. Which of the following is largest?
  a. 0.45
  b. 0.096
  c. 0.3
  d. 0.313

31. What is the value of b in this equation?
$$5b - 4 = 2b + 17$$

  a. 13
  b. 24
  c. 7
  d. 21

32. A school has 15 teachers and 20 teaching assistants. They have 200 students. What is the ratio of faculty to students?
  a. 3:20
  b. 4:17
  c. 3:2
  d. 7:40

33. Express the solution to the following problem in decimal form:
$$\frac{3}{5} \times \frac{7}{10} \div \frac{1}{2}$$

    a. 0.042
    b. 84%
    c. 0.84
    d. 0.42

34. A student gets an 85% on a test with 20 questions. How many answers did the student solve correctly?
    a. 15
    b. 16
    c. 17
    d. 18

35. If Sarah reads at an average rate of 21 pages in four nights, how long will it take her to read 140 pages?
    a. 6 nights
    b. 26 nights
    c. 8 nights
    d. 27 nights

36. Alan currently weighs 200 pounds, but he wants to lose weight to get down to 175 pounds. What is this difference in kilograms? (1 pound is approximately equal to 0.45 kilograms.)
    a. 9 kg
    b. 11.25 kg
    c. 78.75 kg
    d. 90 kg

37. Johnny earns $2334.50 from his job each month. He pays $1437 for monthly expenses. Johnny is planning a vacation in 3 months' time that he estimates will cost $1750 total. How much will Johnny have left over from three months' of saving once he pays for his vacation?
    a. $948.50
    b. $584.50
    c. $852.50
    d. $942.50

38. What is $\frac{420}{98}$ rounded to the nearest integer?
    a. 3
    b. 4
    c. 5
    d. 6

39. Solve the following:

$$3 \times 8 - (2 + 1)^2 + 8$$

    a. 23
    b. 7
    c. 24
    d. 17

40. The total perimeter of a rectangle is 36 cm. If the length of each side is 12 cm, what is the width?
    a. 3 cm
    b. 12 cm
    c. 6 cm
    d. 8 cm

41. Dwayne has received the following scores on his math tests: 78, 92, 83, 97. What score must Dwayne get on his next math test to have an overall average of at least 90?
    a. 89
    b. 98
    c. 95
    d. 100

42. What is the overall median of Dwayne's current scores: 78, 92, 83, 97?
    a. 19
    b. 85
    c. 83
    d. 87.5

43. Solve the following:

$$\left(\sqrt{36} \times \sqrt{16}\right) - 3^2$$

    a. 30
    b. 21
    c. 15
    d. 13

44. In Jim's school, there are 3 girls for every 2 boys. There are 650 students in total. Using this information, how many students are girls?
    a. 260
    b. 130
    c. 65
    d. 390

45. What is the solution to $4 \times 7 + (25 - 21)^2 \div 2$?
    a. 512
    b. 36
    c. 60.5
    d. 22

46. Kimberley earns $10 an hour babysitting, and after 10 p.m., she earns $12 an hour, with the amount paid being rounded to the nearest hour accordingly. On her last job, she worked from 5:30 p.m. to 11 p.m. In total, how much did Kimberley earn on her last job?
    a. $45
    b. $57
    c. $62
    d. $42

47. Solve this equation:

$$9x + x - 7 = 16 + 2x$$

    a. $x = -4$

    b. $x = 3$

    c. $x = \frac{9}{8}$

    d. $x = \frac{23}{8}$

48. Arrange the following numbers from least to greatest value:
    $0.85, \frac{4}{5}, \frac{2}{3}, \frac{91}{100}$

    a. $0.85, \frac{4}{5}, \frac{2}{3}, \frac{91}{100}$

    b. $\frac{4}{5}, 0.85, \frac{91}{100}, \frac{2}{3}$

    c. $\frac{2}{3}, \frac{4}{5}, 0.85, \frac{91}{100}$

    d. $0.85, \frac{91}{100}, \frac{4}{5}, \frac{2}{3}$

49. Keith's bakery had 252 customers go through its doors last week. This week, that number increased to 378. Express this increase as a percentage.
    a. 26%
    b. 50%
    c. 35%
    d. 12%

50. If $4x - 3 = 5$, then $x =$
    a. 1
    b. 2
    c. 3
    d. 4

Answer Explanations

1. D: $\frac{3}{100}$. Each digit to the left of the decimal point represents a higher multiple of 10 and each digit to the right of the decimal point represents a quotient of a higher multiple of 10 for the divisor. The first digit to the right of the decimal point is equal to the value ÷ 10. The second digit to the right of the decimal point is equal to the value ÷ (10 × 10), or the value ÷ 100.

2. B: 13,078. The power of 10 by which a digit is multiplied corresponds with the number of zeros following the digit when expressing its value in standard form. Therefore:

$$(1 \times 10^4) + (3 \times 10^3) + (7 \times 10^1) + (8 \times 10^0)$$

$$10,000 + 3,000 + 70 + 8 = 13,078$$

3. A: 847.90. The hundredths place value is located two digits to the right of the decimal point (the digit 9). The digit to the right of the desired place value is examined to decide whether to round up or keep the digit. In this case, the digit in the thousandths place is examined. Because it is a 6, which is 5 or greater, the hundredths place is rounded up. When rounding up, if the digit to be increased is a 9, the digit to its left is increased by one and the digit in the desired place value is made a zero. Therefore, the number is rounded to 847.90.

4. B: $\frac{11}{15}$. Fractions must have like denominators to be added. The least common multiple of the denominators 3 and 5 is found. The LCM is 15, so both fractions should be changed to equivalent fractions with a denominator of 15. To determine the numerator of the new fraction, the old numerator is multiplied by the same number by which the old denominator is multiplied to obtain the new denominator. For the fraction $\frac{1}{3}$, 3 multiplied by 5 will produce 15. Therefore, the numerator is multiplied by 5 to produce the new numerator $\left(\frac{1\times5}{3\times5} = \frac{5}{15}\right)$. For the fraction $\frac{2}{5}$, multiplying both the numerator and denominator by 3 produces $\frac{6}{15}$. When fractions have like denominators, they are added by adding the numerators and keeping the denominator the same:

$$\frac{5}{15} + \frac{6}{15} = \frac{11}{15}$$

5. C: 34. When performing calculations consisting of more than one operation, the order of operations should be followed: *Parenthesis, Exponents, Multiplication/Division, Addition/Subtraction*.

Parenthesis: $7^2 - 3 \times (4 + 2) + 15 \div 5 = 7^2 - 3 \times (6) + 15 \div 5$.

Exponents: $7^2 - 3 \times 6 + 15 \div 5 = 49 - 3 \times 6 + 15 \div 5$.

Multiplication/Division (from left to right): $49 - 3 \times 6 + 15 \div 5 = 49 - 18 + 3$.

Addition/Subtraction (from left to right): $49 - 18 + 3 = 34$.

6. D: 80%. To convert a fraction to a percent, the fraction is first converted to a decimal. To do so, the numerator is divided by the denominator: $4 \div 5 = 0.8$. To convert a decimal to a percent, the number is multiplied by 100: $0.8 \times 100 = 80\%$.

7. C: 80 min. To solve the problem, a proportion is written consisting of ratios comparing distance and time. One way to set up the proportion is: $\frac{3}{48} = \frac{5}{x}$ $\left(\frac{distance}{time} = \frac{distance}{time}\right)$ where x represents the unknown value of time. To solve a proportion, the ratios are cross-multiplied:

$$(3)(x) = (5)(48) \rightarrow 3x = 240$$

The equation is solved by isolating the variable, or dividing by 3 on both sides, to produce $x = 80$.

8. D: 1, 2, 3, 4, 6, 12. A given number divides evenly by each of its factors to produce an integer (no decimals). The number 5, 7, 8, 9, 10, 11 (and their opposites) do not divide evenly into 12. Therefore, these numbers are not factors.

9. A: 11. To determine the number of houses that can fit on the street, the length of the street is divided by the width of each house: $345 \div 30 = 11.5$. Although the mathematical calculation of 11.5 is correct, this answer is not reasonable. Half of a house cannot be built, so the company will need to either build 11 or 12 houses. Since the width of 12 houses (360 feet) will extend past the length of the street, only 11 houses can be built.

10. B: $3x^2 - 3x + 11$. By distributing the implied 1 in front of the first set of parentheses and the -1 in front of the second set of parentheses, the parenthesis can be eliminated:

$$1(5x^2 - 3x + 4) - 1(2x^2 - 7) = 5x^2 - 3x + 4 - 2x^2 + 7.$$

Next, like terms (same variables with same exponents) are combined by adding the coefficients and keeping the variables and their powers the same:

$$5x^2 - 3x + 4 - 2x^2 + 7 = 3x^2 - 3x + 11.$$

11. B: 180 miles. The rate, 60 miles per hour, and time, 3 hours, are given for the scenario. To determine the distance traveled, the given values for the rate (r) and time (t) are substituted into the distance formula and evaluated:

$$d = r \times t$$

$$d = \left(\frac{60mi}{h}\right) \times (3h)$$

$$d = 180mi$$

12. D: $x \leq -5$. When solving a linear equation or inequality:

Distribution is performed if necessary:

$$-3(x + 4) \rightarrow -3x{-}12 \geq x + 8$$

This means that any like terms on the same side of the equation/inequality are combined.

The equation/inequality is manipulated to get the variable on one side. In this case, subtracting x from both sides produces $-4x{-}12 \geq 8$.

The variable is isolated using inverse operations to undo addition/subtraction. Adding 12 to both sides produces $-4x \geq 20$.

The variable is isolated using inverse operations to undo multiplication/division. Remember if dividing by a negative number, the relationship of the inequality reverses, so the sign is flipped. In this case, dividing by -4 on both sides produces $x \leq -5$.

13. C: $y = 40x + 300$. In this scenario, the variables are the number of sales and Karen's weekly pay. The weekly pay depends on the number of sales. Therefore, weekly pay is the dependent variable (y) and the number of sales is the independent variable (x). Each pair of values from the table can be written as an ordered pair (x, y): (2,380), (7,580), (4,460), (8,620). The ordered pairs can be substituted into the equations to see which creates true statements (both sides equal) for each pair. Even if one ordered pair produces equal values for a given equation, the other three ordered pairs must be checked. The only equation which is true for all four ordered pairs is $y = 40x + 300$:

$$380 = 40(2) + 300 \rightarrow 380 = 380$$

$$580 = 40(7) + 300 \rightarrow 580 = 580$$

$$460 = 40(4) + 300 \rightarrow 460 = 460$$

$$620 = 40(8) + 300 \rightarrow 620 = 620$$

14. D: $x \geq 1$. The closed dot on one indicates that the value is included in the set. The arrow pointing right indicates that numbers greater than one (numbers get larger to the right) are included in the set. Therefore, the set includes numbers greater than or equal to one, which can be written as $x \geq 1$.

15. A: ○. The core of the pattern consists of 4 items: ▲○○□. Therefore, the core repeats in multiples of 4, with the pattern starting over on the next step. The closest multiple of 4 to 42 is 40. Step 40 is the end of the core (□), so step 41 will start the core over (▲) and step 42 is ○.

16. D: The two lines are neither parallel nor perpendicular. Parallel lines will never intersect or meet. Therefore, the lines are not parallel. Perpendicular lines intersect to form a right angle (90°). Although the lines intersect, they do not form a right angle, which is usually indicated with a box at the intersection point. Therefore, the lines are not perpendicular.

17. B: Cone. A polygon is a closed two-dimensional figure consisting of three or more sides. A decagon is a polygon with 10 sides. A triangle is a polygon with three sides. A rhombus is a polygon with four sides. A cone is a three-dimensional figure and is classified as a solid.

18. C: 374.04. The formula for finding the area of a regular polygon is $A = \frac{1}{2} \times a \times P$, where a is the length of the apothem (from the center to any side at a right angle) and P is the perimeter of the figure. The apothem, a, is given as 10.39 and the perimeter can be found by multiplying the length of one side by the number of sides (since the polygon is regular):

$$P = 12 \times 6 \rightarrow P = 72$$

To find the area, substitute the values for a and P into the formula:

$$A = \frac{1}{2} \times a \times P$$

$$A = \frac{1}{2} \times (10.39) \times (72)$$

$$A = 374.04$$

19. C: 216cm. Because area is a two-dimensional measurement, the dimensions are multiplied by a scale that is squared to determine the scale of the corresponding areas. The dimensions of the rectangle are multiplied by a scale of 3. Therefore, the area is multiplied by a scale of 3^2 (which is equal to 9): $24\ cm \times 9 = 216\ cm$.

20. A: (-3, 2). The coordinates of a point are written as an ordered pair (x, y). To determine the x-coordinate, a line is traced directly above or below the point until reaching the x-axis. This step notes the value on the x-axis. In this case, the x-coordinate is -3. To determine the y-coordinate, a line is traced directly to the right or left of the point until reaching the y-axis, which notes the value on the y-axis. In this case, the y-coordinate is 2. Therefore, the ordered pair is written (-3, 2).

21. C: Perimeter is found by calculating the sum of all sides of the polygon. $9 + 9 + 9 + 8 + 8 + s = 56$, where s is the missing side length. Therefore, 43 plus the missing side length is equal to 56. Thus, the missing side length is 13 cm.

22. C: $\frac{1}{3}$ of the shirts sold were patterned. Therefore, $1 - \frac{1}{3} = \frac{2}{3}$ of the shirts sold were solid. Anytime "of" a quantity appears in a word problem, multiplication should be used. Therefore:

$$192 \times \frac{2}{3} = \frac{192 \times 2}{3} = \frac{384}{3} = 128 \text{ solid shirts were sold}$$

The entire expression is $192 \times \left(1 - \frac{1}{3}\right)$.

23. A: Mean. An outlier is a data value that is either far above or far below the majority of values in a sample set. The mean is the average of all the values in the set. In a small sample set, a very high or very low number could drastically change the average of the data points. Outliers will have no more of an effect on the median (the middle value when arranged from lowest to highest) than any other value above or below the median. If the same outlier does not repeat, outliers will have no effect on the mode (value that repeats most often).

24. C: Line graph. The scenario involves data consisting of two variables, month, and stock value. Box plots display data consisting of values for one variable. Therefore, a box plot is not an appropriate choice. Both line plots and circle graphs are used to display frequencies within categorical data. Neither can be used for the given scenario. Line graphs display two numerical variables on a coordinate grid and show trends among the variables.

25. D: $\frac{1}{12}$. The probability of picking the winner of the race is:

$$\frac{1}{4} \left(\frac{number\ of\ favorable\ outcomes}{number\ of\ tota\ outcomes} \right)$$

Assuming the winner was picked on the first selection, three horses remain from which to choose the runner-up (these are dependent events). Therefore, the probability of picking the runner-up is $\frac{1}{3}$. To determine the probability of multiple events, the probability of each event is multiplied:

$$\frac{1}{4} \times \frac{1}{3} = \frac{1}{12}$$

26. A: Compare each numeral after the decimal point to figure out which overall number is greatest. Answers A (1.43785) and C (1.43592) both have the same tenths (4) and hundredths (3). However, the thousandths is greater in answer A (7), so A has the greatest value overall.

27. D: By grouping the four numbers in the answer into factors of the two numbers of the question (6 and 12), it can be determined that (3 x 2) x (4 x 3) = 6 x 12. Alternatively, each of the answer choices could be prime factored or multiplied out and compared to the original value. 6 × 12 has a value of 72 and a prime factorization of $2^3 \times 3^2$. The answer choices respectively have values of 64, 84, 108, 72, and 144 and prime factorizations of 2^6, $2^2 \times 3 \times 7$, $2^2 \times 3^3$, and $2^3 \times 3^2$, so answer D is the correct choice.

28. C: The sum total percentage of a pie chart must equal 100%. Since the CD sales take up less than half of the chart and more than a quarter (25%), it can be determined to be 40% overall. This can also be measured with a protractor. The angle of a circle is 360°. Since 25% of 360 would be 90° and 50% would be 180°, the angle percentage of CD sales falls in between; therefore, it would be answer C.

29. B: Since $850 is the price *after* a 20% discount, $850 represents 80% of the original price. To determine the original price, set up a proportion with the ratio of the sale price (850) to original price (unknown) equal to the ratio of sale percentage:

$$\frac{850}{x} = \frac{80}{100}$$

(where x represents the unknown original price)

To solve a proportion, cross-multiply the numerators and denominators and set the products equal to each other: (850)(100) = (80)(x). Multiplying each side results in the equation 85,000 = 80x.

To solve for x, divide both sides by 80: $\frac{85,000}{80} = \frac{80}{80}$, resulting in x = 1062.5. Remember that x represents the original price. Subtracting the sale price from the original price ($1062.50 − $850) indicates that Frank saved $212.50.

30. A: To figure out which is largest, look at the first non-zero digits. Answer B's first nonzero digit is in the hundredths place. The other three all have nonzero digits in the tenths place, so it must be A, C, or D. Of these, A has the largest first non-zero digit.

31. C: To solve for the value of b, both sides of the equation need to be equalized.

Start by cancelling out the lower value of -4 by adding 4 to both sides:

$$5b - 4 = 2b + 17$$

$$5b - 4 + 4 = 2b + 17 + 4$$

$$5b = 2b + 21$$

The variable b is the same on each side, so subtract the smaller one ($2b$) from each side:

$$5b = 2b + 21$$

$$5b - 2b = 2b + 21 - 2b$$

$$3b = 21$$

Then divide both sides by 3 to get the value of b:

$$3b = 21$$

$$\frac{3b}{3} = \frac{21}{3}$$

$$b = 7$$

32. D: The total faculty is 15 + 20 = 35. So, the ratio is 35:200. Then, divide both of these numbers by 5, since 5 is a common factor to both, with a result of 7:40.

33. C: The first step in solving this problem is expressing the result in fraction form. Separate this problem first by solving the division operation of the last two fractions. When dividing one fraction by another, invert or flip the second fraction and then multiply the numerator and denominator.

$$\frac{7}{10} \times \frac{2}{1} = \frac{14}{10}$$

Next, multiply the first fraction with this value:

$$\frac{3}{5} \times \frac{14}{10} = \frac{42}{50}$$

Decimals are expressions of 1 or 100%, so multiply both the numerator and denominator by 2 to get the fraction as an expression of 100.

$$\frac{42}{50} \times \frac{2}{2} = \frac{84}{100}$$

In decimal form, this would be expressed as 0.84.

34. C: 85% of a number means that number should be multiplied by 0.85: $0.85 \times 20 = \frac{85}{100} \times \frac{20}{1}$, which can be simplified to:

$$\frac{17}{20} \times \frac{20}{1} = 17$$

The answer is C.

35. D: This problem can be solved by setting up a proportion involving the given information and the unknown value. The proportion is:

$$\frac{21 \; pages}{4 \; nights} = \frac{140 \; pages}{x \; nights}$$

Solving the proportion by cross-multiplying, the equation becomes $21x = 4 \times 140$, where $x = 26.67$. Since this is not an exact number of nights, the answer is rounded up to 27 nights. Twenty-six nights would not give Sarah enough time.

36. B: Using the conversion rate, multiply the projected weight loss of 25 lb by 0.45 $\frac{kg}{lb}$ to get the amount in kilograms (11.25 kg).

37. D: First, subtract $1437 from $2334.50 to find Johnny's monthly savings; this equals $897.50. Then, multiply this amount by 3 to find out how much he will have (in three months) before he pays for his vacation: this equals $2692.50. Finally, subtract the cost of the vacation ($1750) from this amount to find how much Johnny will have left: $942.50.

38. B: Dividing by 98 can be approximated by dividing by 100, which would mean shifting the decimal point of the numerator to the left by 2. The result is 4.2, which rounds to 4.

39. A: To solve this correctly, keep in mind the order of operations with the mnemonic PEMDAS (Please Excuse My Dear Aunt Sally). This stands for Parentheses, Exponents, Multiplication, Division, Addition, Subtraction. Taking it step by step, solve the parentheses first:

$$3 \times 8 - (3)^2 + 8$$

Then, apply the exponent:

$$3 \times 8 - 9 + 8$$

Multiplication is performed next:

$$24 - 9 + 8$$

Subtraction and addition are performed after that:

$$15 + 8 = 23$$

40. C: The formula for the perimeter of a rectangle is P = 2L + 2W, where P is the perimeter, L is the length, and W is the width. The first step is to substitute all of the data into the formula:

$$36 = 2(12) + 2W$$

Simplify by multiplying 2 x 12:

$$36 = 24 + 2W$$

Simplifying this further by subtracting 24 on each side, which gives:

$$36 - 24 = 24 - 24 + 2W$$

$$12 = 2W$$

Divide by 2:

$$6 = W$$

The width is 6 cm. Remember to test this answer by substituting this value into the original formula:

$$36 = 2(12) + 2(6)$$

41. D: To find the average of a set of values, add the values together and then divide by the total number of values. In this case, include the unknown value of what Dwayne needs to score on his next test, in order to solve it.

$$\frac{78 + 92 + 83 + 97 + x}{5} = 90$$

Add the unknown value to the new average total, which is 5. Then multiply each side by 5 to simplify the equation, resulting in:

$$78 + 92 + 83 + 97 + x = 450$$

$$350 + x = 450$$

$$x = 100$$

Dwayne would need to get a perfect score of 100 in order to get an average of at least 90.

Test this answer by substituting back into the original formula.

$$\frac{78 + 92 + 83 + 97 + 100}{5} = 90$$

42. D: For an even number of total values, the *median* is calculated by finding the *mean* or average of the two middle values once all values have been arranged in ascending order from least to greatest. In this case, $(92 + 83) \div 2$ would equal the median 87.5, which is answer *D*.

43. C: Follow the *order of operations* in order to solve this problem. Solve the parentheses first, and then follow the remainder as usual.

$$(6 \times 4) - 9$$

This equals $24 - 9$ or 15.

44. D: Three girls for every two boys can be expressed as a ratio: 3:2. This can be visualized as splitting the school into 5 groups: 3 girl groups and 2 boy groups. The number of students that are in each group can be found by dividing the total number of students by 5:

650 divided by 5 equals 1 part, or 130 students per group

To find the total number of girls, multiply the number of students per group (130) by the number of girl groups in the school (3). This equals 390.

45. B: To solve this correctly, keep in mind the order of operations with the mnemonic PEMDAS (Please Excuse My Dear Aunt Sally). This stands for Parentheses, Exponents, Multiplication, Division, Addition, Subtraction. Taking it step by step, solve the parentheses first:

$$4 \times 7 + (4)^2 \div 2$$

Then, apply the exponent:

$$4 \times 7 + 16 \div 2$$

Multiplication and division are both performed next:

$$28 + 8 = 36$$

46. C: Kimberley worked 4.5 hours at the rate of $10/h and 1 hour at the rate of $12/h. The problem states that her pay is rounded to the nearest hour, so the 4.5 hours would round up to 5 hours at the rate of $10/h.

$$(5h)\left(\frac{\$10}{h}\right) + (1h)\left(\frac{\$12}{h}\right) = \$50 + \$12 = \$62$$

47. D:

$9x + x - 7 = 16 + 2x$	Combine $9x$ and x.
$10x - 7 = 16 + 2x$	
$10x - 7 + 7 = 16 + 2x + 7$	Add 7 to both sides to remove (-7).
$10x = 23 + 2x$	
$10x - 2x = 23 + 2x - 2x$	Subtract 2x from both sides to move it to the other side of the equation.
$8x = 23$	
$\dfrac{8x}{8} = \dfrac{23}{8}$	Divide by 8 to get x by itself.
$x = \dfrac{23}{8}$	

48. C: The first step is to depict each number using decimals. $\frac{91}{100} = 0.91$

Multiplying both the numerator and denominator of $\frac{4}{5}$ by 20 makes it $\frac{80}{100}$ or 0.80; the closest approximation of $\frac{2}{3}$ would be $\frac{66}{100}$ or 0.66 recurring. Rearrange each expression in ascending order, as found in answer C.

49. B: First, calculate the difference between the larger value and the smaller value.

378 − 252 = 126

To calculate this difference as a percentage of the original value, and thus calculate the percentage *increase*, divide 126 by 252, then multiply by 100 to reach the percentage = 50%, answer *B*.

50. B: Add 3 to both sides to get $4x = 8$. Then divide both sides by 4 to get $x = 2$.

Clerical

The clerical skills section on civil service exams is used to evaluate applicants' competency at tasks related to record keeping. Examples of tasks that commonly appear on this section include alphabetizing names, completing forms, filing numbers and names, following directions, reading tables, and verifying information for accuracy.

Alphabetizing problems ask test takers to place names and numbers in the proper order. Names of people are first alphabetized by last name. If the last name is identical, the first name is used. If two people have the same last name and first name, they are alphabetized by middle name or initial. Names without a middle name are alphabetized before names with middle names, and initials are alphabetized before full middle names that start with the same letter. For the purposes of alphabetizing, hyphenated names are treated as one word, and apostrophes are ignored. Businesses follow the same alphabetization rules with some exceptions: Articles (*a, an, the*), short prepositions (*in, for, to*) and conjunctions (*and, but, or*) are ignored; abbreviations are spelled out and then alphabetized; and businesses with the same name are alphabetized by location. Alphabetizing numbers depends on the system, so the directions will indicate the relevant rules.

Other problems will provide a form and ask questions about the information it contains. The questions could ask test takers to complete headings or select where new information would be included on the form. Blank headings and information are typically marked by a number. Question prompts might also provide information about a hypothetical situation and ask test takers to answer based on the form.

Filing problems test alphabetizing skills in a different format. The prompt will provide the name of a person, business, or subject, and the answer choices will be the different groups of a filing system (e.g., Aa-De). The directions might provide special rules for the filing system, but if not, the alphabetization rules described above will apply.

Another type of problem will provide directions for a task and then ask about the order of the steps. In addition, the prompt might ask whether the directions can be followed under different circumstances or if some scenario violates the directions. These types of questions are evaluating test takers' attention to detail, reading skills, and logical thinking.

Problems with tables evaluate test takers' ability to interpret and manipulate data. Interpretation questions ask about the information provided, such as a work schedule. Data manipulation questions ask test takers to analyze the data by calculating the average, percentage, or percentage increase. Averages can be calculated by adding the figures together and then dividing by the number of figures. Percentages can be calculated by expressing the two numbers as a fraction, dividing the numerator (top number) by the denominator (bottom number), and then multiplying by 100. Percentage increase can be calculated by finding the difference between two numbers, dividing by the original number, and then multiplying by 100.

Verification problems require test takers to compare names or numbers. The prompt typically includes an address, name, or string of letters and numbers, and the question asks test takers to identify the exact same series of words, letters, or numbers. More than anything else, verification problems are evaluating test takers' attention to detail.

Practice Questions

Questions 1–5: Answer the questions based on the following form. The numbers on the form represent the information contained in the boxes. For example, "Box 1" refers to the requester's legal name.

FORM FOR FREEDOM OF INFORMATION ACT (FOIA) REQUESTS	
To Be Completed by Requester	
Legal Name 1	Home Address 2
Phone Number 3	Email Address 4
Description of Records Requested 5	
Signature and Date 6	
To Be Completed by FOIA Officer	
Legal Name 7	Employee ID Number 8
Department 9	Office Location 10
Final Decision 11	
Signature and Date 12	

1. Where would an FOIA officer find details about the scope and substance of the requested information?
   a. Box 4
   b. Box 5
   c. Box 10
   d. Box 11

2. Which of the following pieces of information would be found in Box 8?
   a. Internal Revenue Service
   b. SD74177691101
   c. johnsmith@email.com
   d. John S. Smith

3. Which of the following pieces of information would be found in Box 11?
    a. The request is denied because the records are classified as confidential and protected under the privacy statute.
    b. The records relate to the Secretary of State's trip to France in 1952.
    c. The records are currently being held at Department of State offices located at Foggy Bottom, Washington, DC.
    d. The records concern American support of France against Vietnam during the First Indochina War.

4. Which of the following boxes does NOT include the requester's contact information?
    a. Box 2
    b. Box 3
    c. Box 4
    d. Box 5

5. Which of the following boxes would most likely read "U.S. Treasury"?
    a. Box 2
    b. Box 3
    c. Box 9
    d. Box 10

Questions 6–15: Answer the following questions by selecting the name that correctly corresponds with the alphabetical position listed in the question's prompt. Numbers are spelled out as words.

6. Which of the following names would be alphabetized first?
    a. Jacob J. Trout
    b. Jason Jacobs Martinez
    c. Julius T. Turner
    d. Jaclyn Piecora

7. Which of the following names would be alphabetized fourth?
    a. Lisa Alice Salazar
    b. Miguel Stanton
    c. Clayton E. Ross
    d. Michael Stanton

8. Which of the following names would be alphabetized second?
    a. Thelma Darling
    b. Jane Doe
    c. J. Doe
    d. Carlos Dominguez-Santos

9. Which of the following names would be alphabetized first?
    a. Walter Hastings
    b. Walter A. Hastings
    c. Franklin Delano Houston
    d. Joseph P. Holland

10. Which of the following names would be alphabetized third?
   a. Alexis Paxton
   b. Alexander Paisley
   c. Alexandra Paulina
   d. Alex Paddington

11. Which of the following names would be alphabetized fourth?
   a. Christina Castellanos
   b. Christine Castellanos
   c. Zephyr Callahan
   d. Richard Allen Carrows

12. Which of the following names would be alphabetized second?
   a. Ronan Michael Myers
   b. Ross Myers
   c. Robert Morton
   d. Rochelle Martinez-Morton

13. Which of the following names would be alphabetized first?
   a. The Prestige Worldwide Media Corp.
   b. Steve's Savory Soups
   c. Sales & Steals Inc.
   d. Ryan & Rickshaw, LLP

14. Which of the following names would be alphabetized third? In this system, numbers are spelled out and then alphabetized.
   a. Scholarly School Supplies Inc.
   b. 2 Bros. Realty Corp.
   c. Baker's Dozen Sweets & Treats
   d. Wally World Resort & Casino

15. Which of the following names would be alphabetized fourth?
   a. Grizzly Mountain Ski Resort (Hunter, NY)
   b. Grizzly Mountain Ski Resort (Denver, CO)
   c. The Gallery of Modern Art & History
   d. Gustav's Gourmet Solutions Inc.

Questions 16–20: Answer the questions based on the following table. A calculator may be used for these questions.

Drug Overdose Deaths in the United States (2010–2017)*		
Year	Deaths Caused by All Drugs	Deaths Caused by Opioid Drugs
2010	38,329	21,089
2011	41,340	22,784
2012	41,502	23,166
2013	43,982	25,052
2014	47,055	28,647
2015	52,404	33,091
2016	63,632	42,249
2017	70,237	47,600

*Data source: Centers for Disease Control and Prevention, www.cdc.gov/nchs/data/databriefs/db329_tables-508.pdf

16. Deaths caused by all drugs increased the least between which years?
    a. 2011–2012
    b. 2012–2013
    c. 2013–2014
    d. 2014–2015

17. How many more deaths were caused by opioid drugs in 2017 compared to 2012?
    a. 17,047 deaths
    b. 22,634 deaths
    c. 24,434 deaths
    d. 25,586 deaths

18. What was the approximate percentage increase in deaths caused by all drugs between 2010 and 2017?
    a. 40%
    b. 45%
    c. 83%
    d. 85%

19. What was the average number of deaths caused by opioid drugs between 2015 and 2017?
    a. 34,366 deaths
    b. 39,955 deaths
    c. 40,980 deaths
    d. 42,420 deaths

20. What percentage of all drug-related deaths was caused by opioid drug use in 2017?
    a. Approximately 63%
    b. Approximately 68%
    c. Approximately 73%
    d. Approximately 78%

Questions 21–25: Answer the questions based on the following set of directions for a U.S. General Services Administration system manager.

> Directions: The system manager is tasked with making a final decision on whether the U.S. General Services Administration (GSA) will amend the records held about a federal employee. In executing this duty, the system manager must consult legal counsel, agency officials, and the GSA Privacy Act officer. The final decision is based on a comparison between the existing record and the amendment proposed by the employee. The system manager must consider how the amendment would impact the accuracy, relevance, appropriateness, and comprehensiveness of the employee's records. In addition, any amendment must comply with all relevant statutes and regulations.
>
> Upon receiving the request, the system manager has ten workdays to approve or deny the proposed amendment in writing. If denied, the decision must be accompanied with an explanation for the denial and information about the appeal process, all in writing. In the event of an administrative delay, the system manager must notify the employee applicant in writing about the delay and state an estimated time frame within ten days of receiving the initial request.

21. What are the time constraints on responding to an amendment request?
    a. System managers must respond to amendment requests within five days of receiving it.
    b. System managers must respond to amendment requests within ten days of receiving it.
    c. System managers must respond to amendment requests within fifteen days of receiving it.
    d. System managers must respond to amendment requests within twenty days of receiving it.

22. Which of the following is the most accurate summary of what a system manager must do if there's an administrative delay?
    a. The system manager must notify the employee applicant about the delay immediately after receiving the request.
    b. The system manager must consider how the delay impacts the amendment's accuracy.
    c. The system manager must provide the employee with information about the appeals process.
    d. The system manager must give written notice to the employee applicant and provide a timetable within ten days of receiving the request.

23. Which of the following best explains what the system manager must consider when deciding whether to approve or deny an amendment?
    a. The system manager must follow the agency officials' final decision.
    b. The system manager must solely evaluate an amendment request based on its compliance with every relevant statute and regulation.
    c. The system manager must accept any amendment that doesn't violate the privacy of any employee or legal entity.
    d. The system manager must analyze the amendment's accuracy, appropriateness, comprehensiveness, legal compliance, and relevance.

24. What event triggers the start of the ten-day deadline for a system manager's response to the employee applicant?
    a. The ten-day deadline starts when an administrative delay is discovered.
    b. The ten-day deadline starts when the agency initiates the review process.
    c. The ten-day deadline starts when the system manager receives the request.
    d. The ten-day deadline starts when the employee applicant first proposes the amendment.

25. Who is responsible for making final decisions on amendment requests?
    a. Legal counsel
    b. System manager
    c. GSA Privacy Act officer
    d. Secretary of the U.S. General Services Administration

Questions 26–30: Answer the questions by selecting the correct filing group for the subject or name that appears inside the quotation marks. The filing system is alphabetical, and it follows standard rules for alphabetization. Numbers are spelled out and then alphabetized accordingly.

26. Which of the following groups would "Top Secret Security Clearance" be filed under?
    a. Tal-Tea
    b. Teb-Tim
    c. Tin-Tub
    d. Tuc-Ube

27. Which of the following groups would "The Reading Project Group" be filed under?
    a. Aaa-Fur
    b. Fus-Lla
    c. Llb-Pur
    d. Pus-Zzz

28. Which of the following groups would "Ignatius W. McNally " be filed under?
    a. Fuz-Gaf
    b. Gag-Kal
    c. Kam-Lup
    d. Luq-Nab

29. Which of the following groups would "Michael Patrick Berry-Smith" be filed under?
    a. Aaa-Fig
    b. Fih-Lug
    c. Luh-Paw
    d. Pax-Zzz

30. Which of the following groups would "Annual Budgetary Allocations" be filed under?
    a. Aar-Abe
    b. Abf-All
    c. Alm-Amm
    d. Amn-Arg

Questions 31–35: Answer the questions based on the following set of directions for Federal Bureau of Investigation (FBI) special agents.

Directions: Federal Bureau of Investigation (FBI) special agents must have a reasonable suspicion of criminal activity before opening an investigation. In addition, the criminal activity must include a violation of federal law for the FBI to have jurisdiction. While conducting their initial investigation, FBI special agents may request search warrants from judges based on probable cause. Once a criminal suspect has been identified during the course of their investigation, FBI special agents may obtain an arrest warrant from a judge by showing probable cause, secure an indictment from a grand jury, or arrest the suspect without obtaining a warrant or indictment. If the suspect is arrested without an arrest warrant or grand jury indictment, the prosecutor must bring the suspect before a judge and show that a crime has been committed based on probable cause within seventy-two hours of the arrest.

31. What must happen before a criminal suspect is arrested?
    a. The FBI must conduct an initial investigation into the suspected federal criminal activity.
    b. The FBI must obtain a search warrant from a grand jury.
    c. The FBI must obtain an arrest warrant from a judge based on probable cause.
    d. The FBI must continue gathering information until the criminal suspect is convicted in a court of law.

32. Who provides search warrants to special agents?
    a. Agency official
    b. Grand juries
    c. Judges
    d. Prosecutor

33. Search warrants are granted based on what legal standard?
    a. Beyond a reasonable doubt
    b. Clear and convincing
    c. Probable cause
    d. Reasonable suspicion

34. What must the prosecutor do after a special agent arrests a criminal suspect without a warrant or indictment?
    a. The prosecutor must prove there is a reasonable suspicion that the suspect violated federal law and not merely state law.
    b. The prosecutor must obtain a search warrant from a judge to complete the investigation within seventy-two hours of the arrest.
    c. The prosecutor must bring the suspect before a grand jury to secure an indictment.
    d. The prosecutor must bring the suspect before a judge and meet the probable cause standard within seventy-two hours of the arrest.

35. What do special agents need prior to opening an investigation?
    a. Special agents must have a reasonable suspicion of federal criminal activity.
    b. Special agents must prove the suspect's guilt based on probable cause.
    c. Special agents must request authorization from a judge.
    d. Special agents must obtain permission to conduct an investigation from a grand jury.

Questions 36–45: Answer the following questions by choosing the name of the person, business, or office code that is spelled exactly like the prompt. The exact spelling includes everything inside the quotation marks.

36. Which of the following choices is spelled exactly like "Jeffrey & Johnson Co."?
    a. Jeffry & Johnson Co.
    b. Jeffrey & Johnston Co.
    c. Jeffrey & Johnson Corp.
    d. Jeffrey & Johnson Co.

37. Which of the following choices is spelled exactly like "Alejandra E. Ramirez"?
    a. Alejandra Ramirez
    b. Alexandra E. Ramirez
    c. Alejandra E. Ramires
    d. Alejandra E. Ramirez

38. Which of the following choices is spelled exactly like "#WM1190EAH3911"?
    a. #WM1190EAH9311
    b. #WM1190EAH3911
    c. #WM1190AEH3911
    d. #WM1190EHA3911

39. Which of the following choices is spelled exactly like "Jackson McHugh"?
    a. Jackson McHugh
    b. Jackson Hugh
    c. Jacksonn McHugh
    d. Jackson M. McHugh

40. Which of the following choices is spelled exactly like "#JD3312RPMM711"?
    a. #JD3312RPMN711
    b. #JD3313RPMM711
    c. #JD3312PPMM711
    d. #JD3312RPMM711

41. Which of the following choices is spelled exactly like "Stationhouse Café & Cabaret"?
    a. Stationhouse Cafe & Cabaret
    b. Stationhouse Café & Cabaret
    c. Station house Café & Cabaret
    d. Stationhouse Cafè & Cabaret

42. Which of the following choices is spelled exactly like "Javier De la Sol"?
    a. Javier De Sol
    b. Javier De la Sal
    c. Javier De la Sol
    d. Javier De La Sol

43. Which of the following choices is spelled exactly like "#LJB8139997A"?
    a. #LJB8139997A
    b. #LJB8133997A
    c. #LJ88139997A
    d. #LIB8139997A

44. Which of the following choices is spelled exactly like "Mats Frederic Lundgren"?
    a. Mats Frederick Lundgren
    b. Matts Frederic Lundgren
    c. Mats Frederic Lundgrenn
    d. Mats Frederic Lundgren

45. Which of the following choices is spelled exactly like "Lucky Louie's Lounge"?
    a. Lucky Louis' Lounge
    b. Lucky Louie's Lounge
    c. Lucky Louie Lounge
    d. Lucky Louis's Lounge

Questions 46–50: Answer the questions based on the following two tables.

Stenographer Courtroom Assignments for January 2019 (Monday–Friday)			
Courtroom	Morning Shift (8 a.m.–noon)	Afternoon Shift (12 p.m.–4 p.m.)	Evening Shift (4 p.m.–8 p.m.)
A	Group 1	Group 1	Group 3
B	Group 4	Group 2	Group 4
C	Group 3	Group 3	Group 2
D	Group 2	Group 4	Group 1

Stenographer Group Assignments for January 2019	
Group 1	Brian, Janet, Karen, Larry
Group 2	Gary, Georgia, Lisa, Mark
Group 3	Ace, Brianna, Xavier, Zephyr
Group 4	Benjamin, Cecilia, David, Paul

46. Who is assigned to the evening shift in Courtroom D?
    a. David
    b. Gary
    c. Larry
    d. Zephyr

47. Who is assigned to the morning shift in Courtroom C?
    a. Brianna
    b. Janet
    c. Karen
    d. Paul

48. Who is never assigned to Courtroom A?
    a. Benjamin
    b. Brian
    c. Larry
    d. Xavier

49. Who is assigned to two shifts in Courtroom B?
    a. Gary
    b. Janet
    c. Paul
    d. Xavier

50. Who is assigned to two shifts in Courtroom C?
    a. Ace
    b. Georgia
    c. Larry
    d. Paul

Answer Explanations

1. B: Details about the request's scope and substance can be found in Box 5, which contains a description of the requested records. Choice *A* is incorrect because Box 4 contains the requester's email address. Choice *C* is incorrect because Box 10 contains the FOIA officer's office location. Choice *D* is incorrect because Box 11 contains the FOIA officer's final decision.

2. B: Box 8 contains the FOIA officer's employee ID number, and Choice *B* appears to be some type of identifying number. Choice *A* is incorrect because "Internal Revenue Service" wouldn't make sense as an employee ID. Choice *C* is incorrect because it is an email address. Lastly, Choice *D* is incorrect because it is a person's name, not an ID number.

3. A: Box 11 is completed by the FOIA officer and labeled "Final Decision." Choice *A* says the request has been denied because the records are classified and protected; this expresses a final decision. Choices *B* and *D* are both incorrect because they describe the requested record's substantive content, and that information would be included in Box 5. Likewise, Choice *C* is incorrect because it describes the record's possible location, which would also be included in Box 5, not Box 11.

4. D: Although the requester similarly fills out Box 5, it contains a description of the records requested, which isn't related to contact information. Information pertaining to the requester contact information can be found in Boxes, 2, 3, and 4. Box 2 contains the requester's home address, Box 3 contains the requester's phone number, and Box 4 contains the requester's email address. As such, Choices *A*, *B*, and *C* must be incorrect.

5. C: Box 9 contains the FOIA officer's department, and "U.S. Treasury" seems to indicate some sort of government institution, agency, or department. Choice *A* is incorrect because Box 2 contains the requester's home address. Choice *B* is incorrect because Box 3 contains the requester's phone number. Choice *D* is incorrect because Box 10 is the FOIA officer's office location.

6. B: This question includes fairly simple last names, which are the first basis for alphabetization. All of the choices' first names begin with "J," but that's irrelevant because all of the last names are different and easily alphabetized. From first to last, the choices are alphabetized in the following order: Jason Jacobs Martinez (Choice *B*), Jaclyn Piecora (Choice *D*), Jacob J. Trout (Choice *A*), and Julius T. Turner (Choice *C*). The question asks for the first name, so Choice *B* is correct.

7. B: Alphabetizing these names is tricky because Choices *B* and *D* have the same last name. So, those two choices need to be alphabetized by first name, which is the second basis for alphabetizing names. From first to last, the choices are alphabetized in the following order: Clayton E. Ross (Choice *C*), Lisa Alice Salazar (Choice *A*), Michael Stanton (Choice *D*), and Miguel Stanton (Choice *B*). The question asks for the fourth name, so Choice *B* is correct.

8. C: This list of names requires applying two different special rules for alphabetization. First, when alphabetizing names that are identical except one uses an initial and the other spells out the full name, the name with the initial is alphabetized first. Second, compound names are alphabetized by the first letter of the first word, and the hyphen is ignored. From first to last, the choices are alphabetized in the following order: Thelma Darling (Choice *A*), J. Doe (Choice *C*), Jane Doe (Choice *B*), and Carlos Dominguez-Santos (Choice *D*). The question asks for the second name, so Choice *C* is correct.

9. A: Although Choices *A* and *B* have the same last name, the one without a middle initial is alphabetized first. All of the other names are alphabetized by last name. From first to last, the choices are alphabetized in the following order: Walter Hastings (Choice *A*), Walter A. Hastings (Choice *B*), Joseph P. Holland (Choice *D*), and Franklin Delano Houston (Choice *C*). The question asks for the first name, so Choice *A* is correct.

10. C: All of the names have a variation of the same first name, but that is a red herring. The list can simply be alphabetized by last name because they are all different. From first to last, the choices are alphabetized in the following order: Alex Paddington (Choice *D*), Alexander Paisley (Choice *B*), Alexandra Paulina (Choice *C*), and Alexis Patton (Choice *A*). The question asks for the third name, so Choice *C* is correct.

11. B: Choices *A* and *B* have the same last name, so they are alphabetized by first name. Those choices' first names are also very similar, but they can be alphabetized by the difference in the final letter. From first to last, the choices are alphabetized in the following order: Zephyr Callahan (Choice *C*), Richard Allen Carrows (Choice *D*), Christina Castellanos (Choice *A*), and Christine Castellanos (Choice *B*). The question asks for the fourth name, so Choice *B* is correct.

12. C: Choices *A* and *B* have the same last name, so they are alphabetized by first name. In addition, Choice *D* has a compound last name, meaning it is alphabetized by the first letter of the first word, and the hyphen is ignored. From first to last, the choices are alphabetized in the following order: Rochelle Martinez-Morton (Choice *D*), Robert Morton (Choice *C*), Ronan Michael Myers (Choice *A*), and Ross Myers (Choice *B*). The question asks for the second name, so Choice *C* is correct.

13. A: Businesses are alphabetized based on very similar rules as people's names, with some slight exceptions. In this question, "The" in Choice *A* is ignored because it is an article, and the rest are alphabetized based on the first letter of the first word. From first to last, the choices are alphabetized in the following order: The Prestige Worldwide Media Corp. (Choice *A*), Ryan & Rickshaw, LLP (Choice *D*), Sales & Steals Inc. (Choice *C*), and Steve's Savory Soups (Choice *B*). The question asks for the first name, so Choice *A* is correct.

14. B: The directions indicate that numbers are spelled out and then alphabetized, so Choice *B* should be alphabetized as "Two Bros. Realty Corp." From first to last, the choices are alphabetized in the following order: Baker's Dozen Sweets & Treats (Choice *C*), Scholarly School Supplies Inc. (Choice *A*), 2 Bros. Realty Corp. (Choice *B*), and Wally World Resort & Casino (Choice *D*). The question asks for the third name, so Choice *B* is correct.

15. D: When businesses have the same name with identical spelling (Choices *A* and *B*), they are alphabetized based on location. In addition, "The" in Choice *D* is ignored because it is an article. From first to last, the choices are alphabetized in the following order: The Gallery of Modern Art & History (Choice *C*), Grizzly Mountain Ski Resort (Denver, CO) (Choice *B*), Grizzly Mountain Ski Resort (Hunter, NY) (Choice *A*), and Gustav's Gourmet Solutions Inc. (Choice *D*). The question asks for the fourth name, so Choice *D* is correct.

16. A: The question is asking for the smallest increase in deaths caused by all drugs, which is calculated by subtracting the years. However, this question can be answered without an exact calculation. The years 2011 and 2012 (Choice *A*) both have death totals in the 41,000s, and the rest of the answer choices are differentiated by several thousand deaths. Alternatively, the correct answer could be found by doing the calculation for every choice. There were 162 more deaths in 2012 than 2011 (Choice *A*); 2,480 more deaths in 2013 than 2012; 3,073 more deaths in 2014 than 2013; and 5,349 more deaths in

2015 than 2014. Thus, Choice *A* is the correct answer because it's the smallest increase in deaths caused by all drugs.

17. C: The question is asking how many more deaths were caused by opioid drugs in 2017 compared to 2012. So, the number of opioid-related deaths in 2012 (23,166) is subtracted from the number of opioid-related deaths in 2017 (47,600), resulting in a difference of 24,434 deaths. Choices *A*, *B*, and *D* incorrectly calculate the difference in opioid-related deaths between 2017 and 2012.

18. C: The question is asking for the percentage increase in deaths caused by all drugs between 2010 and 2017. Percentage increase is calculated by subtracting two numbers, dividing by the original number, and then multiplying by 100. First, the total number of deaths in 2010 (38,329) is subtracted from the total number of deaths in 2017 (70,237), resulting in a difference of 31,908. Second, that difference is divided by the 2010 death total, resulting in a quotient of .832. Third, the quotient is multiplied by 100, resulting in the final answer of 83.2%. Since the question asks for an approximate answer that means Choice *C* is the correct answer. Choices *A*, *B*, and *D* incorrectly calculate the percentage increase in deaths caused by all drugs between 2010 and 2017.

19. C: The question is asking for the average number of deaths caused by opioid drugs between 2015 and 2017. Averages are calculated by adding the figures and then dividing by the number of figures. The opioid-related death totals for this period are 33,091, 42,249, and 47,600, resulting in a cumulative total of 122,940. That total is then divided by 3 because 3 figures were added together, resulting in an average of 40,980 deaths, so Choice *C* is the correct answer. Choices *A*, *B*, and *D* incorrectly calculate the average number of deaths caused by opioid drugs between 2015 and 2017.

20. B: The question is asking for the percentage of all drug-related deaths caused by opioid drugs in 2017. Percentages are calculated by dividing opioid-related deaths by all drug-related deaths and then multiplying by 100. Opioid-related deaths in 2017 were 47,600, and all drug-related deaths were 70,237. Dividing the two numbers results in a quotient of .6777, and the quotient is then multiplied by 100, resulting in the final answer of 67.77%. Choice *B* expresses the same approximate percentage after rounding to the nearest percentage point, so it is the correct answer. Choices *A*, *C*, and *D* incorrectly calculate the percentage of all drug-related deaths caused by opioid drugs in 2017.

21. B: The system manager always needs to provide written notice within ten days of receiving a request regardless of whether it's an approval, denial, or notice of an administrative delay. The other choices are incorrect because the directions do not mention a five-day deadline (Choice *A*), fifteen-day deadline (Choice *C*), or twenty-day deadline (Choice *D*).

22. D: When the system manager knows an administrative delay will prevent the final decision from being made within ten days of receiving the request, the system manager must notify the employee applicant in writing within ten days of receiving the request. This notice must provide an explanation for the delay and state an estimated timetable. Choice *A* is incorrect because the timetable to provide notice is not immediate; it's ten days from receiving the request, and that notice must also be accompanied with a new timetable. Choice *B* is incorrect because an administrative delay isn't related to the amendment's accuracy, which is a factor in the final decision. Similarly, Choice *C* is incorrect because the appeals process occurs after the final decision, so it is irrelevant to an administrative delay.

23. D: The final decision is based on a comparison between the existing record and the amendment, and in making this decision, the system manager must consider the amendment's impact on the record's accuracy, appropriateness, comprehensiveness, legal compliance, and relevance. Thus, Choice *D* is the correct answer. Choice *A* is incorrect because the system manager makes the final decision, not agency officials. Choice *B* is incorrect because legal compliance is only one of several factors that are considered. Choice *C* is incorrect because, although amendments must be in accordance with privacy statutes, that compliance does not automatically result in an amendment being accepted.

24. C: The ten-day deadline starts when the system manager receives the request. Choice *A* is incorrect because notice of an administrative delay must be provided within ten days of receiving the request, not discovery of the delay. Similarly, Choices *B* and *D* are both incorrect because the deadline is not triggered by the agency initiating a review or the employee applicant sending a proposal.

25. B: The system manager is responsible for making the final decision on amendment requests. Choices *A* and *C* are both incorrect because legal counsel and the GSA Privacy Act officer are consulted during the review process, but they do not make the final decision. Choice *D* is also incorrect because the Secretary of the U.S. General Services Administration is not mentioned in the directions.

26. C: The groups are filed alphabetically. So, "Top Secret Security Clearance" would be filed under "Top," which fits between "Tin" and "Tub." Choices *A* and *B* are both incorrect because those groups contain information filed before "Top Secret Security Clearance" alphabetically. Choice *D* is incorrect because it contains information filed after "Top Secret Security Clearance" alphabetically.

27. D: The filing system follows the standard rules for alphabetization. As such, "The Reading Project Group" would be filed under "Reading" because "The" is an article. "Reading" fits between "Pus" and "Zzz." Choices *A*, *B*, and *C* are all incorrect because those groups contain information filed before "The Reading Project Group" alphabetically.

28. D: The filing system follows the standard rules for alphabetization, so "Ignatius W. McNally" is alphabetized based on the last name. "McNally" fits between "Luq" and "Nab." Thus, Choice *D* is correct. Choices *A*, *B*, and *C* are all incorrect because those groups contain information filed before "Ignatius W. McNally" alphabetically.

29. A: The filing system follows the standard rules for alphabetization, so "Michael Patrick Berry-Smith" is filed based on the first letter of the compound last name. "Berry" fits between "Aaa" and "Fig." Choices *B*, *C*, and *D* are all incorrect because those groups contain information filed after "Michael Patrick Berry-Smith" alphabetically.

30. D: The filing system follows the standard rules for alphabetization. Accordingly, "Annual Budgetary Allocations" is filed under the first word because it isn't an article, short preposition, or conjunction. "Annual" fits between "Amn and "Arg." Choices *A*, *B*, and *C* are all incorrect because those groups contain information filed before "Annual Budgetary Allocations" alphabetically.

31. A: The FBI must have a reasonable suspicion of federal criminal activity and conduct an initial investigation into that activity. The investigation must precede the arrest because the investigation is how the FBI identifies the suspect. Choice *B* is incorrect because, although search warrants can help lead to an arrest, they are not required. Choice *C* is incorrect because an arrest warrant is not required for an arrest. Special agents can arrest a suspect without a warrant or obtain a grand jury indictment. Choice *D* is incorrect because the court case takes place after the arrest.

32. C: During the course of their investigation, FBI special agents may request search warrants from judges based on probable cause. Choice *A* is incorrect because the directions never mention agency officials. Choice *B* is incorrect because, according to the directions, grand juries only provide indictments to arrest suspects. Choice *D* is incorrect because prosecutors are responsible for bringing the suspect before a judge within seventy-two hours of a warrantless arrest.

33. C: Search warrants and arrest warrants are both granted based on the probable cause legal standard. Choice *A* is incorrect because, although reasonable suspicion is the legal standard for conducting an investigation, it is unrelated to search warrants. Choices *B* and *D* are both incorrect because, even though they are legal standards, they are not mentioned in the directions.

34. D: If a special agent arrests a criminal suspect without a warrant or indictment, a prosecutor must bring the suspect before a judge within seventy-two hours of the arrest. Once in front of the judge, the prosecutor must meet the probable cause legal standard. Choice *A* is incorrect because the legal standard isn't reasonable suspicion. Choice *B* is incorrect because the investigation doesn't need to be completed within seventy-two hours of the arrest. Choice *C* is incorrect because it misstates the prosecutor's responsibilities during a warrantless arrest.

35. A: Special agents not only need a reasonable suspicion of criminal activity before opening an investigation but the criminal activity must also violate federal law. Choice *B* is incorrect because the special agent isn't responsible for proving the suspect's guilt before opening an investigation. The investigation is how the special agent builds a case to eventually prove the suspect's guilt. Choices *C* and *D* are both incorrect because judges and grand juries don't authorize or provide permission for investigations.

36. D: Choice *D* is identical to the business name provided in the prompt. They both read "Jeffrey & Johnson Co." Choice *A* is incorrect because it is missing an "e" in Jeffrey. Choice *B* is incorrect because it includes a "t" in Johnson. Choice *C* is incorrect because it lists the business as "Corp." instead of "Co."

37. D: Choice *D* is identical to the name of the person provided in the prompt. They both read "Alejandra E. Ramirez." Choice *A* is incorrect because it is missing the middle initial. Choice *B* is incorrect because it replaces the "j" with an "x" in the first name. Choice *C* is also incorrect because it spells the last name with an "s" instead of a "z."

38. B: Choice *B* is identical to the code provided in the prompt. They both read "#WM1190EAH3911." Choice *A* is incorrect because it reverses the order of the "3" and "9." Similarly, Choice *C* is incorrect because it reverses the order of the "E" and "A," and Choice *D* is incorrect because it reverses the order of the "A" and "H."

39. A: Choice *A* is identical to the name of the person provided in the prompt. They both read "Jackson McHugh." Choice *B* is incorrect because it's missing the "Mc" in the last name. Likewise, Choice *C* is incorrect because it adds an extra "n" to Jackson. Choice *D* is incorrect because it adds a middle initial to the name.

40. D: Choice *D* is identical to the code provided in the prompt. They both read "#JD3312RPMM711." Choice *A* is incorrect because it switches the "MM" in the code to "MN." Choice *B* is incorrect because it changes the sequence from "3312" to "3313." Choice *C* is incorrect because it switches the "RP" to "PP."

41. B: Choice *B* is identical to the business name provided in the prompt. They both read "Stationhouse Café & Cabaret." Choice *A* is incorrect because it spells "Café" without an accent mark. Choice *C* is incorrect because it spells "Stationhouse" as two words. Choice *D* is incorrect because the accent mark in "Café" is reversed.

42. C: Choice *C* is identical to the name of the person provided in the prompt. They both read "Javier De la Sol." Choice *A* is incorrect because it's missing "la." Choice *B* is incorrect because "Sol" is spelled with an "a" instead of an "o." Choice *D* is incorrect because it capitalizes "la."

43. A: Choice *A* is identical to the code provided in the prompt. They both read "#LJB8139997A." Choice *B* is incorrect because it changes the sequence from "139997" to "133997." Choice *C* is incorrect because it changes "B8" to "88." Choice *D* is incorrect because it switches the "J" to "I" at the beginning of the code.

44. D: Choice *D* is identical to the name of the person provided in the prompt. They both read "Mats Frederic Lundgren." Choice *A* is incorrect because it spells "Frederic" with a "k." Choice *B* is incorrect because it spells "Mats" with an extra "t." Choice *C* is incorrect because it spells "Lundgren" with an extra "n."

45. B: Choice *B* is identical to the business name provided in the prompt. They both read "Lucky Louie's Lounge." Choice *A* is incorrect because it spells "Louie" as "Louis." Choice *C* is incorrect because it's missing the apostrophe and "s" in "Louie's." Choice *D* is incorrect because it spells "Louie" with an "s" instead of an "e."

46. C: Group 1 is assigned to the evening shift in Courtroom D. As a result, the correct answer must be a member of Group 1 (Larry). Choices *A*, *B*, and *D* are all incorrect because none of those stenographers work the evening shift in Courtroom D. David is a member of Group 4, Gary is a member of Group 2, and Zephyr is a member of Group 3.

47. A: Group 3 is assigned to the morning shift in Courtroom C, so the correct answer will be a member of that group (Brianna). Choices *B* and *C* are both incorrect because Janet and Karen are members of Group 1. Choice *D* is incorrect because Paul is a member of Group 4.

48. A: Group 1 is assigned to the morning and afternoon shifts in Courtroom A, and Group 3 is assigned to the evening shift in Courtroom A. So, the correct answer can be a member of either Group 2 or Group 4 because both groups never work in Courtroom A. Benjamin is a member of Group 4. Choices *B* and *C* are incorrect because Brian and Larry are members of Group 1. Choice *D* is incorrect because Xavier is a member of Group 3.

49. C: Group 4 is assigned to the morning and evening shifts in Courtroom B, and no other group is assigned to two shifts in that courtroom. Consequently, the correct answer will be a member of Group 4 (Paul). Choices *A*, *B*, and *D* are all incorrect because none of those stenographers work two shifts in Courtroom B. Gary is a member of Group 2, Janet is a member of Group 1, and Xavier is a member of Group 3.

50. A: Group 3 is assigned to the morning and afternoon shifts in Courtroom C, and no other group is assigned to two shifts in that courtroom. As such, the correct answer must be a member of Group 3 (Ace). Choices *B*, *C*, and *D* are all incorrect because none of those stenographers work two shifts in Courtroom C. Georgia is a member of Group 2, Larry is a member of Group 1, and Paul is a member of Group 4.

Dear Civil Service Test Taker,

We would like to start by thanking you for purchasing this study guide for your exam. We hope that we exceeded your expectations.

Our goal in creating this study guide was to cover all of the topics that you will see on the test. We also strove to make our practice questions as similar as possible to what you will encounter on test day. With that being said, if you found something that you feel was not up to your standards, please send us an email and let us know.

We have study guides in a wide variety of fields. If you're interested in one, try searching for it on Amazon or send us an email.

Thanks Again and Happy Testing!
Product Development Team
info@studyguideteam.com

Interested in buying more than 10 copies of our product? Contact us about bulk discounts:
bulkorders@studyguideteam.com

FREE Test Taking Tips DVD Offer

To help us better serve you, we have developed a Test Taking Tips DVD that we would like to give you for FREE. **This DVD covers world-class test taking tips that you can use to be even more successful when you are taking your test.**

All that we ask is that you email us your feedback about your study guide. Please let us know what you thought about it – whether that is good, bad or indifferent.

To get your **FREE Test Taking Tips DVD**, email freedvd@studyguideteam.com with "FREE DVD" in the subject line and the following information in the body of the email:

  a. The title of your study guide.

  b. Your product rating on a scale of 1-5, with 5 being the highest rating.

  c. Your feedback about the study guide. What did you think of it?

  d. Your full name and shipping address to send your free DVD.

If you have any questions or concerns, please don't hesitate to contact us at freedvd@studyguideteam.com.

Thanks again!

H. 6/20

CPSIA information can be obtained
at www.ICGtesting.com
Printed in the USA
LVHW011010140520
654604LV00017B/215

9 781628 456479